AF388669

The 2nd International Online Conference on Polymer Science—Polymers and Nanotechnology for Industry 4.0

The 2nd International Online Conference on Polymer Science—Polymers and Nanotechnology for Industry 4.0

Editors

Gianluca Cicala
Ana María Díez-Pascual
Shin-ichi Yusa

MDPI • Basel • Beijing • Wuhan • Barcelona • Belgrade • Manchester • Tokyo • Cluj • Tianjin

Editors
Gianluca Cicala
University of Catania
Italy

Ana María Díez-Pascual
Universidad de Alcalá
Spain

Shin-ichi Yusa
University of Hyogo
Japan

Editorial Office
MDPI
St. Alban-Anlage 66
4052 Basel, Switzerland

This is a reprint of articles from the Proceedings published online in the open access journal *Materials Proceedings* (ISSN 2673-4605) (available at: https://www.mdpi.com/2673-4605/7/1).

For citation purposes, cite each article independently as indicated on the article page online and as indicated below:

LastName, A.A.; LastName, B.B.; LastName, C.C. Article Title. *Journal Name* **Year**, *Volume Number*, Page Range.

ISBN 978-3-0365-5287-3 (Hbk)
ISBN 978-3-0365-5288-0 (PDF)

Cover image courtesy of MDPI

Contents

Tejashree Amberkar and Prakash Mahanwar
Synthesis and Study of Microcapsules with Beeswax Core and Phenol-Formaldehyde Shell
Using the Taguchi Method [†]
Reprinted from: *Mater. Proc.* **2021**, 7, 1, doi:10.3390/IOCPS2021-11207 1

Kalyani Pathak, Ratna Jyoti Das, Riya Saikia, Aparoop Das and Mohammad Zaki Ahmad
Bora Rice: Natural Polymer for Drug Delivery [†]
Reprinted from: *Mater. Proc.* **2021**, 7, 2, doi:10.3390/IOCPS2021-11290 7

Ahmed Fatimi
Hydrogel-Based Bioinks for Three-Dimensional Bioprinting: Patent Analysis [†]
Reprinted from: *Mater. Proc.* **2021**, 7, 3, doi:10.3390/IOCPS2021-11239 13

Bin Jeremiah D. Barba, David P. Peñaloza, Jr., Noriaki Seko and Jordan F. Madrid
RAFT-Mediated Radiation Grafting on Natural Fibers in Aqueous Emulsion [†]
Reprinted from: *Mater. Proc.* **2021**, 7, 4, doi:10.3390/IOCPS2021-11243 21

**Bin Jeremiah D. Barba, Patricia Nyn L. Heruela, Patrick Jay E. Cabalar, John Andrew A. Luna,
Allan Christopher C. Yago and Jordan F. Madrid**
Nanografting of Polymer Brushes on Gold Substrate by RAFT-RIGP [†]
Reprinted from: *Mater. Proc.* **2021**, 7, 5, doi:10.3390/IOCPS2021-11587 29

**Magdi H. Mussa, Yaqub Rahaq, Sarra Takita, Farah D. Zahoor, Nicholas Farmilo and Oliver
Lewis**
The Influence of Adding a Functionalized Fluoroalkyl Silanes (PFDTES) into a Novel
Silica-Based Hybrid Coating on Corrosion Protection Performance on an
Aluminium 2024-t3 Alloy [†]
Reprinted from: *Mater. Proc.* **2021**, 7, 6, doi:10.3390/IOCPS2021-11240 37

Irene Abelenda Núñez, Ramón G. Rubio, Francisco Ortega and Eduardo Guzmán
Hyaluronic Acid Hydrogel Particles Obtained Using Liposomes as Templates [†]
Reprinted from: *Mater. Proc.* **2021**, 7, 7, doi:10.3390/IOCPS2021-11222 45

**Anne Yagolovich, Andrey Kuskov, Pavel Kulikov, Leily Kurbanova, Anastasia Gileva and
Elena Markvicheva**
Antitumor Cytokine DR5-B-Conjugated Polymeric Poly(N-vinylpyrrolidone) Nanoparticles
with Enhanced Cytotoxicity in Human Colon Carcinoma 3D Cell Spheroids [†]
Reprinted from: *Mater. Proc.* **2021**, 7, 8, doi:10.3390/IOCPS2021-11281 51

Cansu Esen and Baris Kumru
Light-Driven Integration of Graphitic Carbon Nitride into Polymer Materials [†]
Reprinted from: *Mater. Proc.* **2021**, 7, 9, doi:10.3390/IOCPS2021-11590 53

Christos Panagiotopoulos, Dimitrios Korres and Stamatina Vouyiouka
Vitrimerization of Poly(butylene succinate) By Reactive Melt Mixing Using Zn(II)
Epoxy-Vitrimer Chemistry [†]
Reprinted from: *Mater. Proc.* **2021**, 7, 10, doi:10.3390/IOCPS2021-11588 61

**Christina I. Gkountela, Dimitrios N. Markoulakis, Dimitrios M. Korres and Stamatina N.
Vouyiouka**
Evaluation of the Parameters of Poly(Butylene succinate) Enzymatic Polymerization [†]
Reprinted from: *Mater. Proc.* **2021**, 7, 11, doi:10.3390/IOCPS2021-11274 63

Adam Olszewski, Paulina Kosmela, Łukasz Zedler, Krzysztof Formela and Aleksander Hejna
Optimization of Foamed Polyurethane/Ground Tire Rubber Composites Manufacturing [†]
Reprinted from: *Mater. Proc.* **2021**, *7*, 12, doi:10.3390/IOCPS2021-11244 . . . . **65**

Alvin Kier R. Gallardo, Lorna S. Relleve and Alyan P. Silos
Gel Properties of Carboxymethyl Hyaluronic Acid/Polyacrylic Acid Hydrogels Prepared by Electron Beam Irradiation [†]
Reprinted from: *Mater. Proc.* **2021**, *7*, 13, doi:10.3390/IOCPS2021-11220 . . . . **73**

Tejashree Amberkar and Prakash Mahanwar
Review on Thermal Energy Storing Phase Change Material-Polymer Composites in Packaging Applications [†]
Reprinted from: *Mater. Proc.* **2021**, *7*, 14, doi:10.3390/IOCPS2021-11218 . . . . **81**

Chunyin Lu, Jianhui Qiu, Eiichi Sakai and Guohong Zhang
A Novel 3D Microporous Structure Hydrogel with Stable Mechanical Properties and High Elasticity and Its Application in Sensing [†]
Reprinted from: *Mater. Proc.* **2021**, *7*, 15, doi:10.3390/IOCPS2021-11212 . . . . **87**

Antonella Patti, Stefano Acierno, Gianluca Cicala, Mauro Zarrelli and Domenico Acierno
Assessment of Recycled PLA-Based Filament for 3D Printing [†]
Reprinted from: *Mater. Proc.* **2021**, *7*, 16, doi:10.3390/IOCPS2021-11209 . . . . **89**

Tejashree Amberkar and Prakash Mahanwar
Study and Characterization of Phase Change Material-Recycled Paperboard Composite for Thermoregulated Packaging Applications [†]
Reprinted from: *Mater. Proc.* **2021**, *7*, 17, doi:10.3390/IOCPS2021-11208 . . . . **97**

Pieter Samyn, Joey Bosmans and Patrick Cosemans
Current Alternatives for In-Can Preservation of Aqueous Paints: A Review [†]
Reprinted from: *Mater. Proc.* **2021**, *7*, 18, doi:10.3390/IOCPS2021-11245 . . . . **105**

Ahmed Fatimi, Oseweuba Valentine Okoro and Amin Shavandi
Biopolymer-Based Hydrogels for 3D Bioprinting [†]
Reprinted from: *Mater. Proc.* **2021**, *7*, 19, doi:10.3390/IOCPS2021-11284 . . . . **113**

Elena Togliatti, Cosimo C. Laporta, Maria Grimaldi, Olimpia Pitirollo, Antonella Cavazza, Diego Pugliese, Daniel Milanese and Corrado Sciancalepore
Preparation and Characterisation of PBAT-Based Biocomposite Materials Reinforced by Protein Complex Microparticles [†]
Reprinted from: *Mater. Proc.* **2021**, *7*, 20, doi:10.3390/IOCPS2021-12019 . . . . **115**

Konstantina Chronaki, Angeliki Mytara, Constantine D. Papaspyrides, Konstantinos Beltsios and Stamatina Vouyiouka
A Novel Treatment Tool for PLA-Based Encapsulation Systems [†]
Reprinted from: *Mater. Proc.* **2021**, , 21, doi:10.3390/IOCPS2021-11268 . . . . **121**

Lara Velasco Davoise, Rafael Peña Capilla and Ana M. Díez-Pascual
Assessment of the Optical Properties of a Graphene–Poly(3-hexylthiophene) Nanocomposite Applied to Organic Solar Cells [†]
Reprinted from: *Mater. Proc.* **2021**, *7*, 22, doi:10.3390/IOCPS2021-11241 . . . . **123**

Proceeding Paper

Synthesis and Study of Microcapsules with Beeswax Core and Phenol-Formaldehyde Shell Using the Taguchi Method [†]

Tejashree Amberkar * and **Prakash Mahanwar**

Department of Polymer and Surface Engineering, Institute of Chemical Technology, Mumbai 400019, India; pa.mahanawar@ictmumbai.edu.in
* Correspondence: tejuamberkar@yahoo.co.in
† Presented at the 2nd International Online Conference on Polymer Science—Polymers and Nanotechnology for Industry 4.0, 1–15 November 2021; Available online: https://iocps2021.sciforum.net/.

Abstract: Phenol-formaldehyde shelled phase change material microcapsules (MPCMs) were fabricated and their processing parameters were analyzed with the Taguchi method. Core to shell ratio, surfactant concentration and speed of mixing are the parameters that were optimized in five levels. The optimized values for the surfactant concentration, core to shell ratio and agitation speed were 3%, 1:1 and 800 rpm, respectively. The obtained microcapsules were spherical in shape. The melting enthalpy of the MPCMs synthesized with optimized processing parameters was 148.93 J/g in 35–62 °C. The obtained temperature range of phase transition temperature can be used for storing different food articles such as chocolate and hot served foods.

Keywords: phase change material; thermal energy storage; beeswax; latent heat

1. Introduction

Citation: Amberkar, T.; Mahanwar, P. Synthesis and Study of Microcapsules with Beeswax Core and Phenol-Formaldehyde Shell Using the Taguchi Method. *Mater. Proc.* **2021**, *7*, 1. https://doi.org/10.3390/IOCPS2021-11207

Academic Editor: Shin-ichi Yusa

Published: 20 October 2021

Beeswax has been used as a phase change material (PCM) in various applications such as building applications and solar energy storage applications owing to its phase change temperature around 60 °C. Beeswax has been used in waterborne coating for preparing hydrophobic thermoregulating coating [1]. The dispersion of beeswax/perfluorinated copolymer (used as an encapsulant)/silica nanoparticle was made with a homogenizer to give a 350-nanometer particle size. The obtained material had a phase transition enthalpy of 84 J/g at 61 °C. The addition of thermally conductive nanomaterials in a beeswax composite improves its heat dissipation characteristics. Beeswax has been mixed with copper, aluminum and graphite nanoparticles to increase its conductivity [2]. The prepared composite filled in 55-millimeter capsules and was placed in a heat storage tank. The graphite/beeswax nanocomposite outperforms other nanocomposites with less time for charging and a high discharging time. Bentonite clay has been integrated with beeswax and mixed with concrete [3]. The heat absorption of the composite with PCM increased by 6.67%, but the compression strength was reduced. Modified carbon nanotubes (CNTs) (5%) have been incorporated in beeswax using vacuum impregnation [4]. The incorporation of CNTs increased the thermal conductivity. The composite had a reduced melting enthalpy of 115.5 J/g at 60 °C. Black beeswax has been incorporated into a prototype roof model made from an MDF sheet and covered with EPS foam [5]. An increase in the temperature at night of 3.6 °C was observed. A simulation study projected an energy saved of 67%. Beeswax has been stored in a container with a PV panel [6]. It reduced heat waste from the panel and increased the voltage generated.

The microencapsulation method is an exhaustively used technique for PCM shape-stabilization. This technique has been used in a broad spectrum of applications such as building, medicine, electronics, food, etc. Polymeric encapsulation is characterized by high toughness and a good heat transfer property due to the large surface area of the capsules.

In this paper, phenol-formaldehyde shelled PCM microcapsules are fabricated, and their properties are studied.

2. Materials and Methods

Phenol and formaldehyde were purchased from SD Fine chemicals private limited. Beeswax was procured from SRL, Mumbai. Polyvinyl alcohol (PVA) were bought from Loba Chemie. Resorcinol, ammonium chloride and xylene were purchased from Research Lab private limited. Deionized water (DI) was used in all the experimental work.

Beeswax/phenol-formaldehyde core/shell particles were prepared using a suspension polymerization technique. In DI water, 5 wt.% aqueous solutions of PVA was prepared. The solution was mixed by magnetic stirring at 500 rpm until PVA dissolved. Under agitation, 2.1 g of phenol and 0.5 g of ammonium chloride were dissolved in PVA solution for 30 min. The pH of PVA solution was adjusted to 7–8 using an ammonia solution. Different amounts of PCM were added to 10 mL of xylene in beakers and subjected to agitation for 5 min with a magnetic stirrer at 60 °C. An emulsion was allowed to form by adding PVA solution to PCM solution under ultrasonication for 30 min. To another heated container placed inside the heater, 3.35 g of 37 wt.% aqueous solution of formaldehyde and ultrasonicated solution was added. Solution was slowly added to the container and maintained at 65 °C under stirring at 500 rpm for the next 2 h. Then, 5 wt.% of HCl was added to maintain the pH at about 3–4 and 0.5 g of resorcinol was added. The reaction was continued at the same temperature for the next 2.5 h. Microcapsules were recovered by filtration under vacuum. The microcapsules were rinsed with water, washed with xylene, and dried for 24 h.

An accurately weighed sample was crushed and stirred in xylene for 1 h at 70 °C under magnetic stirring. The microcapsules were rinsed with water, washed with xylene, and dried for 24 h. The core content is the percentage of microcapsule weight difference before and after the treatment. The core content was calculated by taking an average of 3 readings. An Olympus BX41 optical microscope was used to measure the size of a microcapsule. The size of microcapsule was calculated taking mean of 100 readings from Image J software. A differential scanning calorimeter (Shimadzu DSC-60) was used to determine enthalpy and phase transition temperature of microcapsules.

3. Results

Core to shell ratio, surfactant concentration and speed of mixing are the parameters that need to be optimized. This can be achieved thoroughly with the Taguchi method. The three parameters were varied in five levels. The core to shell ratio was varied as 0.5:1, 1:1, 1.5:1, 2:1 and 2.5:1. The surfactant concentration was varied as 1, 2, 3, 4 and 5%. The agitation speed was varied as 400, 600, 800, 1000 and 1200 rpm. The batches studied are shown in Table 1. Varying all the parameters requires fabrication of 5^3 (125) batches. However, with the help of the Taguchi orthogonal array, the optimized parameters can be obtained with only 25 batches. The increase in the core content will increase the thermal energy storage property. Thus, larger-the-better form of analysis was chosen.

Table 1. Taguchi L_{25} orthogonal array with signal to noise ratio (SNR).

Run (Nos.)	Surfactant Concentration (g)	Core-to-Shell Ratio (In Moles)	Agitation Speed (rpm)	Core Content (wt.%)				SNR (dB)
				R1	R2	R3	Average	
1	1	0.5:1	400	59.4	63	61.2	61.2	35.73503
2	1	01:01	600	77.4	68.4	70.2	72	37.14665
3	1	1.5:1	800	67.5	72	71.1	70.2	36.92674
4	1	02:01	1000	52.2	48.6	50.4	50.4	34.04861
5	1	2.5:1	1200	14.4	27	18	19.8	25.9333
6	2	0.5:1	400	75.6	75.6	79.2	76.8	37.70722
7	2	01:01	600	73.8	77.4	81	77.4	37.77482
8	2	1.5:1	800	68.4	68.4	63	66.6	36.46948

Table 1. *Cont.*

Run (Nos.)	Surfactant Concentration (g)	Core-to-Shell Ratio (In Moles)	Agitation Speed (rpm)	Core Content (wt.%)				SNR (dB)
				R1	R2	R3	Average	
9	2	02:01	1000	52.2	57.6	52.2	54	34.64788
10	2	2.5:1	1200	73.8	77.4	81	77.4	37.77482
11	3	0.5:1	400	66.6	72	82.8	73.8	37.36113
12	3	01:01	600	69	81	84	78	37.84189
13	3	1.5:1	800	57	73.5	72	67.5	36.58608
14	3	02:01	1000	67.5	76.5	76.5	73.5	37.32575
15	3	2.5:1	1200	70.5	67.5	73.5	70.5	36.96378
16	4	0.5:1	400	70.5	72.45	73.5	72.15	37.16473
17	4	01:01	600	79.2	79.2	73.8	77.4	37.77482
18	4	1.5:1	800	62.4	66	58.8	62.4	35.90369
19	4	02:01	1000	61.2	58.8	60	60	35.56303
20	4	2.5:1	1200	66	69.6	73.2	69.6	36.85218
21	5	0.5:1	400	48	45	51	48	33.62482
22	5	01:01	600	57	66	61.5	61.5	35.7775
23	5	1.5:1	800	78	74.4	78	76.8	37.70722
24	5	02:01	1000	64.8	66	60	63.6	36.06914
25	5	2.5:1	1200	60	52.5	67.5	60	35.56303

The effect of the parameter values on the core content can be studied with the main effects plot for SN ratios, which is shown in Figure 1.

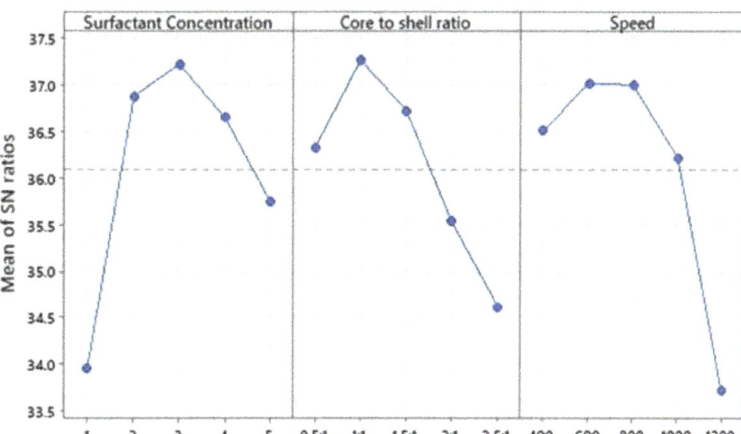

Figure 1. Main effects plots of SN ratio.

Increasing the surfactant concentration gave a finer emulsion with better dispersion. As the surfactant concentration increases above 3 wt.%, the core content reduces. This is the reason for the decrease in the SNR value. An increase in the core content was observed for core to shell ratios 1:1 and 0.5:1. A further increase in the ratio reduced the shell thickness. The ruptured thin shell can show a low core content. An increase in the speed up to 800 rpm helped in the formation of core/shell morphology; increasing the speed above this may rupture the shell. Therefore, the optimized values for the surfactant concentration, core to shell ratio and agitation speed are 3%, 1:1 and 800 rpm.

The size of MPCM with optimized parameters was calculated taking mean of 100 readings in image J software. Thus, the size obtained was 62.61 μm. The optical micrograph was shown in Figure 2. The obtained microcapsules were spherical in shape. The suggested parameters of reaction give small sized microcapsules which can be easily used in coating applications with smaller thickness coatings.

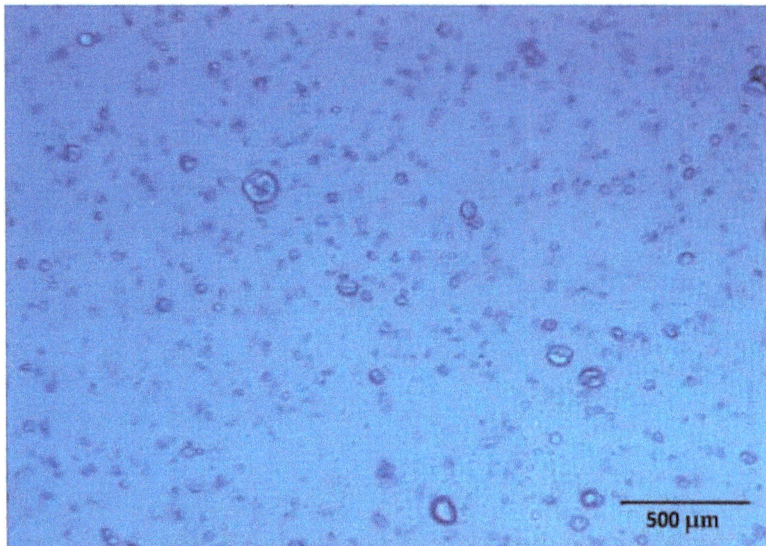

Figure 2. Optical micrograph of optimized batch.

The melting enthalpy of the MPCMs was 148.93 J/g in the range of 35–62 °C. The melting thermogram can be seen in Figure 3. The obtained temperature range of the phase transition temperature can be used for storing different food articles such as chocolate and hot served foods, building materials and solar energy storing materials. The two peaks of the phase transition allow heat storage for a larger temperature range.

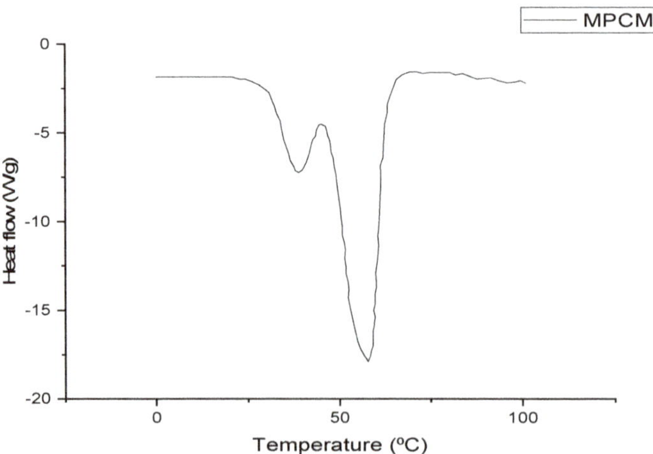

Figure 3. DSC thermogram of optimized batch.

4. Discussion

The effect of the surfactant concentration, core to shell ratio and agitation speed on the core content of MPCMs was studied. The optimized values for the surfactant concentration, core to shell ratio and agitation speed were 3%, 1:1 and 800 rpm. The structure of the MPCM is spherical and the size in micrometers allows it to be used for a myriad of applications. The melting enthalpy and temperature range of the phase transition are suitable for thermal energy storing applications.

Author Contributions: Writing—original draft preparation, T.A.; writing—review and editing, P.M. All authors have read and agreed to the published version of the manuscript.

Funding: This work is funded by AICTE National Doctoral Fellowship Scheme sanctioned vide letter F No.: 12-2/2019-U1 provided by the Ministry of Human Resource Development, Government of India.

Institutional Review Board Statement: Not applicable.

Informed Consent Statement: Not applicable.

Data Availability Statement: The data presented in this study are available on request from the corresponding author.

Acknowledgments: The authors acknowledge research facilities provided by Institute of Chemical Technology, Mumbai.

Conflicts of Interest: The authors declare no conflict of interest.

References

1. Naderizadeh, S.; Heredia-Guerrero, J.A.; Caputo, G.; Grasselli, S.; Malchiodi, A.; Athanassiou, A.; Bayer, I.S. Superhydrophobic Coatings from Beeswax-in-Water Emulsions with Latent Heat Storage Capability. *Adv. Mater. Interfaces* **2019**, *6*, 1801782. [CrossRef]
2. Sravani, V.; Reddy, K.D. Optimisation of Parameters in Thermal Energy Storage System by Enhancing Heat Transfer in Phase Change Material. *IOP Conf. Ser. Earth Environ. Sci.* **2019**, *312*, 012004. [CrossRef]
3. Thaib, R.; Hamdani, H.; Amin, M. Utilization of Beeswax/Bentonite as energy storage material on building wall composite. *J. Phys. Conf. Ser.* **2019**, *1402*, 044038. [CrossRef]
4. Putra, N.; Rawi, S.; Amin, M.; Kusrini, E.; Kosasih, E.A.; Indra Mahlia, T.M. Preparation of beeswax/multi-walled carbon nanotubes as novel shape-stable nanocomposite phase-change material for thermal energy storage. *J. Energy Storage* **2019**, *21*, 32–39. [CrossRef]
5. Salih, T.W.M.; Abdulrehman, M.A.; Al-Moameri, H.H.; Al-Kamal, A.K. Energy saving in Iraq: Waxes as phase change materials for space heating. *AIP Conf. Proc.* **2020**, *2213*, 020141. [CrossRef]
6. Thaib, R.; Umar, H.; Rizal, T.A. Experimental Study of the Use of Phase Change Materials as Cooling Media on Photovoltaic Panels. *Eur. J. Eng. Technol. Res.* **2021**, *6*, 22–26. [CrossRef]

materials
proceedings

Proceeding Paper

Bora Rice: Natural Polymer for Drug Delivery [†]

Kalyani Pathak [1],*, Ratna Jyoti Das [1], Riya Saikia [1], Aparoop Das [1] and Mohammad Zaki Ahmad [2]

[1] Department of Pharmaceutical Sciences, Dibrugarh University, Dibrugarh 786004, India; ratnajyotidas@gmail.com (R.J.D.); saikia.riya27@gmail.com (R.S.); aparoopdas@dibru.ac.in (A.D.)
[2] Department of Pharmaceutics, College of Pharmacy, Najran University, Najran 55461, Saudi Arabia; zaki.manipal@gmail.com
* Correspondence: kalyakster@gmail.com
† Presented at the 2nd International Online Conference on Polymers Science—Polymers and Nanotechnology for Industry 4.0, 1–15 November 2021; Available online: https://iocps2021.sciforum.net/.

Abstract: Natural polymers play a vital part in the formulation of pharmaceutical dosage forms due to their use as excipients. Synthetic polymers have been introduced into drug delivery recently; the usage of natural polymers in drug delivery research continues to rise. It is not surprising that applications other than its caloric value have been found for starch. Various natural sources of the polymer have been investigated for delivery systems; among them, *Assam Bora rice* starch seems to be a promising candidate due to its interesting properties such as being non-toxic, biocompatible, biodegradable, mucoadhesive, and non-immunogenic. *Assam Bora rice*, locally known as *Bora Chaul*, was first introduced in Assam, India, from Thailand or Myanmar by Thai-Ahom, now widely cultivated throughout the Assam. The starch obtained from *Assam Bora* rice is characterized by its high amylopectin content (i.e., >95%) with a branched, waxy polymer which shows physical stability and resistance towards enzymatic action. *Assam Bora rice* starch hydrates and swells in cold water, forming viscous colloidal dispersion or sols responsible for its bioadhesive nature. Moreover, it is degraded by colonic bacteria but remains undigested in the upper GIT. Due to the excellent adhesion and gelling capability, it is often selected as a mucoadhesive matrix in a controlled release drug delivery system. Carboxymethyl *Assam Bora rice* starch has also been applied for SPIONs stabilization and, further, it can effectively bind and load cationic anti-cancer drug molecule, Doxorubicin hydrochloride (DOX), via electrostatic interaction. This article provides a critical assessment of *Assam Bora rice* literature and shows how the rice can be used in many ways, from food additives to drug delivery systems.

Keywords: Assam Bora rice; starch; natural polymer; mucoadhesive agent; drug delivery system

Citation: Pathak, K.; Das, R.J.; Saikia, R.; Das, A.; Ahmad, M.Z. Bora Rice: Natural Polymer for Drug Delivery. *Mater. Proc.* **2021**, 7, 2. https://doi.org/10.3390/IOCPS2021-11290

Academic Editor: Marina Arrieta

Published: 1 November 2021

Publisher's Note: MDPI stays neutral with regard to jurisdictional claims in published maps and institutional affiliations.

1. Introduction

Natural polymers have been widely explored as vehicles for the encapsulation and delivery of drugs and other bioactive compounds, drawing considerable interest. Their main advantages in terms of remarkable biological properties, including controlled enzyme degradation, selective interactions with specific biomolecules, and versatility of modification, enable them to be used for various drug delivery applications. Additionally, due to the reactive groups present in the indigenous biodegradable polymers, additional functional groups may be inserted, endowing the newly produced materials with great functionalities, or changing their physical and chemical properties. Natural polymers include polysaccharides, proteins, peptides, polyesters, and several others. The Drug Delivery System has extensively investigated the natural polymers for their biocompatibility and processability. Polysaccharide and protein-based materials are similar to the extracellular matrix, giving them minimally invasive features [1,2]. The backbones of polymers are also abundant in groups that can be modified, such as amino, carboxyl, and hydroxyl groups. Some natural polymers have shown stronger affinity to cell receptors and govern cellular processes such as adhesion, proliferation, and migration, which could be used to develop more selective and efficient usages. Enzyme-dependent degradation ensures their ability to build stimuli-responsive delivery systems in local locations [3].

Advances in drug delivery have driven the development of novel excipients that are safe, accomplish specialized functions, and directly or indirectly affect the rate and degree of release or absorption. Today, a wide variety of therapeutic excipients derived from plants are commercially available. Numerous investigations on the efficacy of plant-based materials as pharmaceutical excipients have been conducted. Synthetic polymers are toxic, expensive, have negative environmental consequences, require a long time to make, and are not as abundant as naturally occurring polymers. On the other hand, natural polymers are attractive for pharmaceutical applications as they are affordable, readily available, non-toxic, chemically changeable, potentially biodegradable, and biocompatible with few exceptions. The rationale for the rising importance of natural plant-based materials is that they are renewable and can provide a continual supply of raw materials when cultivated or harvested sustainably [3,4]. Plant-based polymers have been studied for their potential application in various pharmaceutical dosage forms, including matrix-controlled systems, film coating agents, buccal films, microspheres, nanoparticles, and viscous liquid formulations such as ophthalmic solutions, suspensions, and implants.

Additionally, these compounds have been employed as stabilizers, disintegrants, solubilizers, emulsifiers, suspending agents, gelling agents, bioadhesives, and binders [5]. Natural polymers are biocompatible and free of unwanted effects as they are essentially polysaccharides. Native starch is a carbohydrate isolated from its botanical source with little processing to retain its intrinsic physicochemical properties following processing. Starch is one of the most popular biopolymers in the drug delivery system due to its versatility as an excipient in drug manufacturing, and its inexpensive cost. Starch possesses a wide variety of intrinsic physical and chemical features that dictate its functional properties and applications [6,7]. Starch can be mainly obtained from crops, including maize (*Zea mays*), rice (*Oryza sativa*), wheat (*Triticum aestivum*), and potato (*Solanum tuberosum, Solanum tuberosum*). Due to the versatility of its applications and low cost, starch is one of the most often utilized biopolymers in drug delivery technology [8].

North-east India, notably Assam, is recognized as a confluence point for rice growing and is blessed with an abundance of rice varieties. Interestingly, starch has been proven to have applications other than caloric value. Numerous natural sources of the polymer have been targeted for use in delivery methods, including Assam Bora rice starch appears to be a better candidate due to its unique qualities, which include non-toxicity, biocompatibility biodegradability, mucoadhesiveness, and non-immunogenicity [9]. Assam Bora rice (*Oryza sativa* L, Japonica variant) is a readily available Assamese glutinous rice with a high amylopectin content. Sticky rice starch is used in a range of pharmaceutical applications. Assam Bora rice, locally known as *Bora Chaul*, was introduced to Assam, India, by Thai-Ahom from Thailand or Myanmar. It is currently widely cultivated throughout Assam. The starch derived from Assam Bora rice is distinguished by a high amylopectin concentration (>95%) and a branched waxy polymer that demonstrates physical stability and resistance to enzymatic action. Assam Bora rice starch hydrates and swells in cold water, generating viscous colloidal dispersion, or sols, that contribute to its bioadhesive properties. Additionally, it is destroyed by colonic bacteria but remains undigested in the upper gastrointestinal tract (GIT). It is frequently used as a mucoadhesive matrix in controlled release medication delivery systems due to its outstanding adherence and gelling properties. Assam Bora rice has a well-documented composition, physicochemical characteristics, shape, and medicinal applications [9,10].

This review focuses on the pharmaceutical applications and issues associated with using this native polymer in drug manufacturing technology, particularly in the formulation of conventional tablets and capsules and in specific controlled drug delivery systems. A breakthrough is likely to occur in using natural polymeric materials if the existing vigorous research on the use of natural polymeric materials is sustained and maintained. This advancement will likely address several disadvantages associated with this class of potential pharmaceutical excipients, thereby altering the landscape of the preferred pharmaceutical excipient for drug delivery in the future.

2. Physicochemical Properties of *Assam Bora Rice*

Bora rice, also known as sticky rice or glutinous rice, is a short-grain Asian variant used in many cuisines for its characteristic sticky texture and possesses an excellent gelling property. There is limited information on the structure and physicochemical properties of *Assam Bora* rice starch. Properties are normal for their biological origin. From previous research, it has been demonstrated that the physicochemical properties of starch from *Assam Bora rice* are pretty similar to those of rice starches. *Assam Bora rice* is white or almost white, odourless, and tasteless. The amount of amylose in the product was nearly negligible. Bora rice starch contains a high concentration of amylopectin. Amylopectin is a complex molecule with three distinct types of branch chains. Bora rice, a rice variety high in amylopectin, has adhesive qualities and can thus be used alone or in conjunction with plant mucilage in appropriate proportions to develop matrix-type drug delivery systems. Due to its origin, *Assam Bora rice* starch has a polygonal to spherical shape and a reasonably smooth surface. The infrared spectrum of starch is almost identical to the conventional infrared spectrum. In 1 M KOH, the inherent viscosity of *Assam Bora rice* starch is much higher than that of other rice starches. Moisture absorption capacity of *Assam Bora rice* starch increases as relative humidity increases. *Assam Bora rice* starch has high crystallinity. Starch granules were revealed to be semicrystalline in nature, and crystallinity has been attributed to the well-ordered structure of amylopectin molecules within the granules, as amylopectin has been shown to alter the crystallinity level of starch granules. The degree of crystallinity of starch granules has an effect on a variety of starch properties, including gelatinization, resistance to hydrolysis (both acid and enzyme), and reactivity during chemical modification. The rigid structure of *Assam Bora rice* starch granules may account for the undigested nature of the Bora rice diet, which retains its bulk and medicinal benefits. As *Assam Bora rice* starch is resilient to enzymatic hydrolysis, it is well suited for colon targeted drug delivery systems [2,4,11,12].

3. Bora Rice as a Natural Polymer

3.1. Drug Delivery Applications

With advancement in science and technology, innovations in the field of public health care and medication conveyance have turned out to be one of the most promising fields for researchers across the globe. Newer designs are developed to establish controlled drug delivery systems that provide beneficial and advantageous therapeutic potency and efficacious release of drugs at the target site compared to the traditional regimens that demand frequent dosing. The use of distinct and appropriate polymers as excipient also has a vital role to play in the development of specific drug delivery systems. The utilization of natural polymers in drug delivery has attracted a lot of attention from researchers worldwide [2]. Natural polymers exhibit various advantages such as cost-effectiveness, enhanced bioavailability, easy accessibility, conservation, and, most importantly, fewer unwanted effects. Bora rice has evolved to be one of the promising natural polymers that are very stable and being designated as 'GRAS' (Generally regarded as safe), which is the primary criteria for any excipient to be used for administrative and pharmaceutical purposes [10,11]. Bora rice also finds its application in medication conveyance, as it has been characterized by some special properties. As it is obtained from natural origin, the margin of safety is much higher when compared to synthetic polymers. However, some of the utilities of Bora rice as excipients in different drug delivery systems have been investigated [11–15].

a.   Bora rice starch as directly compressible agents:

Assam Bora rice starch (ABRS) exhibits the property of a binder and directly compressible agents in the formulation of tablets. In order to study these properties of Bora rice starch, a model medication comprising atorvastatin was used for developing tablets that were prepared under different conditions. The tablets were prepared with acid-altered ABRF, untreated ABRF, thermally treated ABRF, and spray-dried ABRF, which were further analysed, and compared with tablets utilizing microcrystalline cellulose. The study results indicated that thermally and untreated ABRF did not attain the required hardness when prepared without a binding agent. However, spray dried, and acid-modified tablets were used as directly compressible agents. Studies were also performed to evaluate the compression properties, Young's modulus, uniformity in weight, dissolution, tensile strength, friability, toughness, the drug-content profile of ABRFs. In addition, the mechanical properties of ABRFs were analysed using Kawakita and Heckel's method, and the results of the study signified the possible use of ABRFs as directly compressible agents for tablet dosage forms when compared to starch [13,14].

b. Plasma volume expander properties:

The chemical composition of Bora rice starch indicates the presence of amylopectin that is also present in the structure of glycogen, and hence it was studied for its polymer expanding properties. The characterization of the Bora rice starch was the subject to various analytical and other related studies such as 1H NMR, the Mark–Houwink relationship, FTIR studies, the relationship between viscosity and molecular weight, and osmotic pressure utilizing inward estimation technique. The result of the studies indicated that Bora rice starch could be efficiently used as a plasma volume expander [2].

c. Importance of Bora rice in the development of targeted drug delivery system:

Towards the development of targeted drug delivery systems, a series of experimental studies were performed using Bora rice starch. ABRS was used to establish a novel colon-targeted tablet (compression covered) that was meant to target the 5-FU receptor site. Spray-dried lactose and microcrystalline cellulose were used to prepare the core tablet that was further coated with ABRS which played the role of a targeted drug carrier. The study results inferred that ABRS could potentially be used as a drug carrier in targeted delivery systems. Further research into the field also indicated that ABRS exhibits biocompatible mucoadhesive properties utilized in the development of mucoadhesive microspheres (MAMs). To achieve targeted delivery, the MAMs were prepared by means of a double emulsion solvent evaporation system [11–13].

d. Bora rice starch as biopolymeric polymer:

A series of in vitro dissolution studies were performed for the microbeads prepared by ionotropic gelation method following SUPAC-MR rules that incorporated the mixture of sodium alginate and pregelatinized Bora rice. The microbeads were analysed using different parameters such as mucoadhesion, surface morphology, pharmacotechnical boundaries, and drug-polymer compatibilities. Further, the result of the studies indicated that Bora rice starch could be efficiently used as a biopolymeric excipient in drug delivery systems, and in medication transporter frameworks it can also be used as drug release modulators [2,10–15].

3.2. Upcoming Challenges of Bora Rice in Drug Delivery

Even though Bora rice has multiple functional versatilities, there are many properties that lower the efficiency of Bora rice, making it a less dependable pharmaceutical excipient in innovative and conventional formulations. Before selecting an excipient for any particular formulation, it is vital to analyze the basic information associated with that excipient, and this is considered the basis of the final selection. Absorption properties and intrinsic moisture content are vital properties that can provide preliminary information about the excipient. Moisture content can affect properties such as flow, tensile strength, and compaction properties of the tablets and granules. The moisture content of the Bora rice starch is directly associated with the degree of relative humidity (RH) present in the atmosphere and the environment into which the starch material is stored. Thus, the moisture content of Bora rice elevates with an increase in the sorption of moisture when the RH of the environment increases [2]. In the case of Bora rice, the process of moisture sorption occurs in different phases that begin with the tight binding of water molecules to the anhydrous glucose units.

The initial step continues until a stoichiometric ratio of 1:1 (water:anhydrous glucose) is attained throughout the Bora rice starch grains. With an increase in water absorption, the stoichiometric ratio results as 1:2 (water:anhydrous glucose), imparting the qualities of bulk water influencing the compaction and flow properties of starch. From the research studies, it has been observed that there is a sharp fall in the Granules' flowability at 60% RH, and the flowability completely stops at 70% RH.

High moisture content has been found to produce adverse effects on the flow properties of the granules resulting in a serious variation in their weight. Hence, these weight variation issues produce serious issues during the production of capsules, tablets, and granules packaged into sachets, especially when automated systems are employed for large batch production. Furthermore, it has been observed that Bora rice starch forms cake when exposed to high RH. Studies have indicated that the highest strength of tablets can be attained when the ratio of water:anhydrous glucose stoichiometry corresponds to 1:1 and 1:2, considered to be the equilibrium moisture content state. This stoichiometry can be obtained by storing the Bora rice starch at RHs 60% and 70%, respectively. Thus, the concentration of moisture is an essential parameter for powder and granules, primarily when Bora rice starch is employed in high concentrations or when it is used as diluents in the formulation [11–13]. On the other hand, moisture content also has an influential role in the compaction properties of Bora rice starch.

Further, the poor flow property exhibited by the rice starch is also responsible for producing minimal lubricant sensitivity, adding more hindrances to the use of Bora rice in drug delivery. To get rid of these obstacles, science-based data on the formulation, drug-delivery system, potency, dosage efficacy, and the degree of side effects should be considered during the design set of a drug delivery system incorporating Bora rice as an excipient. Hence, it is the prime duty of the researchers to bring a revolution by designing and developing a drug delivery system that can exploit the potential benefits of Bora rice polymer in terms of commercial utilities to overcome the hindrances and challenges of the use of Bora rice in drug delivery systems [15].

4. Conclusions

Native starches are a safe biopolymer with several pharmaceutical applications. They are obtained from botanical sources and processed to pharmaceutical standards. Various natural sources of the polymer have been investigated for delivery systems; among them, *Assam Bora rice* starch seems to be a promising candidate due to its interesting properties such as being non-toxic, biocompatible, biodegradable, mucoadhesive, and non-immunogenic. Starch isolated from *Assam Bora rice* revealed that it exhibits the same kind of physicochemical properties as rice starches. The therapeutic potential of *Assam Bora rice*, as evidenced by its extraordinarily high amylopectin concentration, is presented and studied as a matrix operator for controlled release drug delivery systems. As a result, the remarkable properties of Bora rice can be used to accelerate the development of drug delivery systems. Bora rice starch may be utilized as an excipient in the future to deliver drugs with poor physical and chemical qualities in a controlled/sustained/prolonged manner. Additionally, there is no patent on Bora rice starch, which opens new possibilities for researching the potential commercial benefits of Bora rice starch polymer. Bora rice starch is also being used in nano-sized colloidal, vesicular, and specialized carrier systems. According to the aforementioned literature, the use of Bora rice for nanotechnology may have an additive or synergistic effect on water-insoluble pharmaceuticals' delivery, increasing their physicochemical qualities and changing their pharmacokinetics and pharmacodynamics.

Author Contributions: All the authors have worked collaboratively to add new data collections, theories, and new concepts related to the Assam Bora rice. All authors have read and agreed to the published version of the manuscript.

Funding: This research received no external funding.

Institutional Review Board Statement: Not applicable.

Informed Consent Statement: Not applicable.

Data Availability Statement: Not applicable.

Conflicts of Interest: The authors declare no conflict of interest.

References

1. Das, P.; Singha, A.D.; Goswami, K.; Sarmah, K. Detection of Nutritionally Significant Indigenous Rice Varieties from Assam, India. *BEPLS* **2018**, *7*, 5.
2. Ahmad, M.Z.; Bhattacharya, A. Isolation and Physicochemical Characterization of Assam Bora Rice Starch for Use as a Plasma Volume Expander. *Curr. Drug Deliv.* **2010**, *7*, 162–167. [CrossRef] [PubMed]
3. Sharma, H.K.; Mukherjee, A.; Nath, L.K. Evaluation and Comparison of Treated-Untreated Assam Bora Rice Flour for Use as Directly Compressible Agent. *Int. J. Curr. Biomed. Pharm. Res.* **2011**, *1*, 173–177.
4. Stasiak, K.; Molenda, M.; Opali_nski, I.; Błaszczak, W. Mechanical properties of native maize, wheat, and potato starches. *Czech. J. Food Sci.* **2013**, *31*, 347–354. [CrossRef]
5. Ačkar, Đ.; Babić, J.; Šubarić, D.; Kopjar, M.; Miličević, B. Isolation of Starch from Two Wheat Varieties and Their Modification with Epichlorohydrin. *Carbohydr. Polym.* **2010**, *81*, 76–82. [CrossRef]
6. Kaur, H.; Yadav, S.; Ahuja, M.; Dilbaghi, N. Synthesis, Characterization and Evaluation of Thiolated Tamarind Seed Polysaccharide as a Mucoadhesive Polymer. *Carbohydr. Polym.* **2012**, *90*, 1543–1549. [CrossRef]
7. Banik, R.; Das, P.; Deka, N.; Sarmah, T.C. Ready to Use Ethnic Rice Products of Assam, India: Potential Source of Resistant Starch. *BEPLS* **2018**, *7*, 5.
8. Quodbach, J.; Kleinebudde, P.A. Critical review on tablet disintegration. *Pharm. Dev. Technol.* **2015**, *21*, 763–774. [CrossRef] [PubMed]
9. Sachan, N.K.; Bhattacharya, A. Evaluation of Assam Bora Rice Starch as a Possible Natural Mucoadhesive Polymer in Formulation of Microparticulate Control Drug Delivery Systems. *J. Assam Sci. Soc.* **2006**, *47*, 34–41.
10. *Collection, Characterization and Conservation of Indigenous Rice Varieties of Assam and Meghalaya*; Gene Campaign & North East Centre for Rural Livelihood Research (NECR): Dergaon, India, 2014.
11. Ahmad, M.Z.; Akhter, S.; Anwar, M.; Singh, A.; Ahmad, I.; Ain, M.R.; Jain, G.K.; Khar, R.K.; Ahmad, F.J.; Khar, R.K.; et al. Feasibility of A Assam bora rice starch as a compression coat of 5-Fluorouracil core tablet for colorectal Cancer. *Curr. Drug Deliv.* **2012**, *9*, 105–110. [CrossRef]
12. Ramteke, K.H.; Nath, L. Formulation, Evaluation and Optimization of Pectin- Bora Rice Beads for Colon Targeted Drug Delivery System. *Adv. Pharm. Bull.* **2014**, *4*, 167–177.
13. Santander-Ortega, M.J.; Stauner, T.; Loretz, B.; Ortega-Vinuesa, J.L.; Bastos-González, D.; Wenz, G.; Schaefer, U.F.; Lehr, C.M. Nanoparticles Made from Novel Starch Derivatives for Transdermal Drug Delivery. *J. Control. Release* **2010**, *141*, 85–92. [CrossRef] [PubMed]
14. Shaptadvip, B.; Sarma, R.N. Assessment of Nature and Magnitude of Genetic Diversity Based on DNA Polymorphism with RAPD Technique in Traditional Glutinous Rice (*Oryza sativa* L.) of Assam. *Asian J. Plant Sci.* **2009**, *8*, 218–223. [CrossRef]
15. Tripathi, K.K.; Warrier, R.; Govila, O.P.; Ahuja, V. *Biology of Oryza sativa L. (Rice)*; India: Series of Crop Specific Biology Documents; Ministry of Science and Technology: New Delhi, India, 2011.

materials proceedings

MDPI

Proceeding Paper

Hydrogel-Based Bioinks for Three-Dimensional Bioprinting: Patent Analysis [†]

Ahmed Fatimi [1,2]

1   Department of Chemistry, Polydisciplinary Faculty, Sultan Moulay Slimane University, P.O. Box 592, Mghila, Beni-Mellal 23000, Morocco; a.fatimi@usms.ma
2   ERSIC, Polydisciplinary Faculty, Sultan Moulay Slimane University, P.O. Box 592, Mghila, Beni-Mellal 23000, Morocco
†   Presented at the 2nd International Online Conference on Polymer Science—Polymers and Nanotechnology for Industry 4.0, 1–15 November 2021; Available online: https://iocps2021.sciforum.net/.

Abstract: There are a variety of hydrogel-based bioinks commonly used in three-dimensional bioprinting. In this study, in the form of patent analysis, the state of the art has been reviewed by introducing what has been patented in relation to hydrogel-based bioinks. Furthermore, a detailed analysis of the patentability of the used hydrogels, their preparation methods and their formulations, as well as the 3D bioprinting process using hydrogels, have been provided by determining publication years, jurisdictions, inventors, applicants, owners, and classifications. The classification of patents reveals that most inventions intended for hydrogels used as materials for prostheses or for coating prostheses are characterized by their function or properties Knowledge clusters and expert driving factors show that biomaterials, tissue engineering, and biofabrication research is concentrated in the most patents.

Keywords: bioinks; hydrogels; 3D bioprinting; patentability; patent data

Citation: Fatimi, A. Hydrogel-Based Bioinks for Three-Dimensional Bioprinting: Patent Analysis. *Mater. Proc.* **2021**, *7*, 3. https://doi.org/ 10.3390/IOCPS2021-11239

Academic Editor: Shin-ichi Yusa

Published: 30 October 2021

Publisher's Note: MDPI stays neutral with regard to jurisdictional claims in published maps and institutional affiliations.

1. Introduction

There are a variety of hydrogels commonly used in three-dimensional (3D) bioprinting, which is a process involving the deposition of cell-laden hydrogels (or bioinks) on a substrate (Figure 1). Hydrogels are synthetic matrices made up of a network of hydrophilic polymers that absorb water and/or biological fluids. They can be created from a large number of water-soluble materials, including synthetic polymers (e.g., polyethylene glycol, polylactic acid, etc.), proteins (e.g., collagen, gelatin, etc.), and polysaccharides (e.g., alginate, hyaluronic acid, etc.) [1–4].

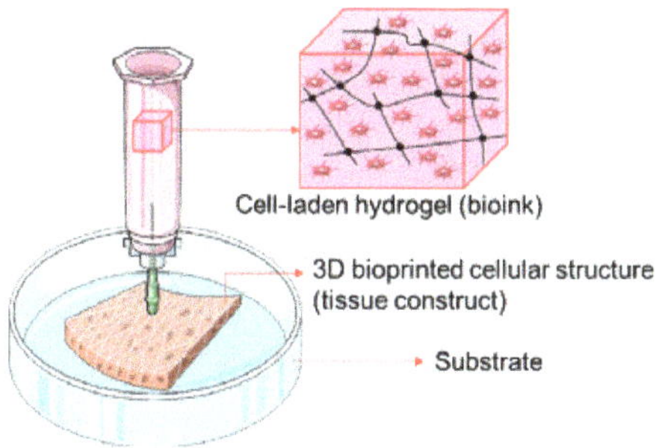

Figure 1. Schema of 3D bioprinting process involving the deposition of cell-laden hydrogels (or bioinks) on a substrate to create 3D cellular structures for tissue engineering applications.

The 3D structure of these hydrogels is due to crosslinking, which forms an insoluble macromolecular network in the environmental fluid (Figure 2). The resemblance to different biological tissues, due to the elasticity and the presence of a large amount of water, allows the use of hydrogels in the regeneration of several types of damaged tissues (e.g., fibrin hydrogel is seeded with neural cells to regenerate brain tissue, keratinocytes are seeded in collagen hydrogel to regenerate skin tissue, etc.) [5,6].

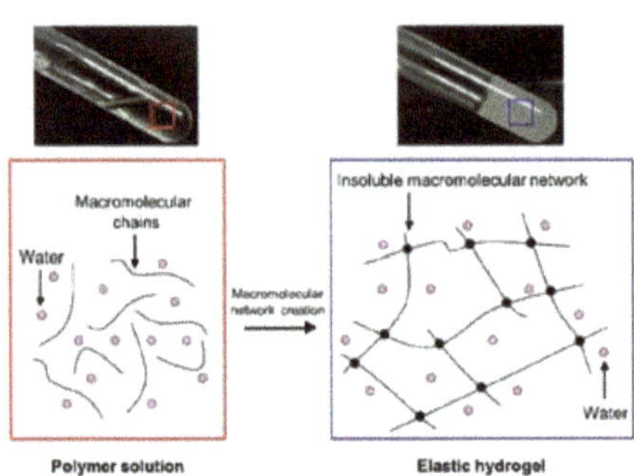

Figure 2. Schema of hydrogel crosslinking, which forms an insoluble macromolecular network in the environmental fluid.

The first patent application concerning hydrogel-based bioinks was filed in 2005, and then granted in 2012 [7]. Through this patent, Forgacs et al. have invented an apparatus comprising, among others, a cartridge comprising one or more cell aggregates. Inventors have then proved the concept by bioprinting a fused ring structure by using cell aggregates and poly(N-(hydroxypropyl)methacrylamide)-based hydrogel as bioink [7].

To promote the sufficiency of bioinks in 3D bioprinting, several researchers have investigated pathways to enhance bioink properties to meet bioprinting requirements, with several hydrogels developed. Research on hydrogels as bioinks is developing rapidly through the innovation and improvement of raw materials (synthetic and natural polymers), synthesis, and methods of preparation, as well as formulations and biofabrication processes. Moreover, more than 100 organizations (universities, academic institutes, companies, foundations, government bodies, etc.) around the world are currently involved in patent activity and filings concerning hydrogel-based bioinks. This trend is justified by the several advantages that hydrogels offer for bioprinting and biomedical applications. This is also evident from the increase in the number of patent applications filed each year worldwide in research and development in this area. For example, patent applications related to hydrogel-based bioinks have increased from 3 to 103, during the period from 2013 to 2020, respectively [8].

This work in the form of patent analysis, which is a family of techniques for studying the information present within and attached to patents, describes the state of the art by introducing what has been innovated and patented in relation to hydrogel-based bioinks regarding to used hydrogels, their preparation methods and their formulations, as well as the 3D bioprinting process using hydrogels. Furthermore, this work gives a competitive analysis of the past, present and future trends in the hydrogel-based bioinks and leads to various recommendations that could help one to plan and innovate research strategy.

2. Resources and Analysis

2.1. Resources and Research Methods

The supported field codes used in this study were based on the Patentscope search service of the World Intellectual Property Organization (WIPO) [8,9] and The Lens patent data set [10]. During the search, different keywords and related terms were used, and patents were searched according to title, abstract, and claims. The search was then filtered to include only documents with an application date until 2020.

2.2. Analysis of the Patentability of Hydrogel-Based Bioinks

The search indicates that 119 patent documents have been found. Generally, it encompasses patent applications and granted patents. In relation to hydrogel-based bioinks, the found patent documents are classed as: 103 patent applications and 16 granted patents.

Here, we will review the state of the art by introducing what has been patented in relation to hydrogel-based bioinks. We then provide a detailed analysis of the patentability of the used hydrogels, their preparation methods and their formulations, as well as the 3D bioprinting process, following these sections:

- Publication year;
- Jurisdictions;
- Inventors;
- Applicants;
- Owners;
- Patent classifications.

3. Results

3.1. Publication Year

The date on which a patent document is published, thereby making it part of the state of the art. For hydrogel-based bioinks, 119 patent documents have been found until 2020. The year 2013 saw the registration of three patent documents only (exclusively patent applications). However, the year 2020 recorded 31 patent documents (24 patent applications and 7 granted patents). The maximum number of patent applications (25) was recorded in 2019. However, the maximum number of granted patents (7) was recorded in 2020.

3.2. Jurisdictions

An applicant, or first mentioned applicant in the case of joint applicants, can file an application for patent at the appropriate Patent Office (e.g., European Patent Office (EPO), United States Patent and Trademark Office (USPTO), Korean Intellectual Property Office (KIPO), China National Intellectual Property Administration (CNIPA), etc.) under whose jurisdiction he normally resides, has his domicile, has a place of business, or the place from where the invention actually originated. For hydrogel-based bioinks, the top 10 jurisdictions of filled patents until 2020 are presented in Table 1.

Table 1. Jurisdictions (top 10) of resulted patents as a function of patent documents and patent contribution (%) of hydrogel-based bioinks.

Jurisdiction	Patent Documents	Patent Contribution (%)
United States	32	26.89
PCT	26	21.85
China	25	21.01
Republic of Korea	16	13.45
Australia	10	8.40
Europe	6	5.04
Sweden	1	0.84
Russia	1	0.84
Eurasia	1	0.84
Canada	1	0.84

The United States through the USPTO encompasses 32 patent documents, with a higher patent contribution per total of ~26.9%. On the other hand, the global system for filing patent applications, known as the Patent Cooperation Treaty (PCT) and administered by WIPO, encompasses 26 patent documents with a patent contribution per total of ~21.8%, as well as China through the CNIPA, encompasses 25 patent documents with a patent contribution per total of ~13.4%. Finally, the EPO, through which patent applications are filed regionally (Europe), encompasses six patent documents with a patent contribution per total of ~5%.

3.3. Inventors

The inventor is a natural person designated for a patent application. In several cases, the inventor can also be the applicant, and there may be more than one inventor per patent application [11].

For hydrogel-based bioinks, the top 10 inventors until 2020 are presented in Table 2. The inventors, Murphy Keith and Dorfman Scott, from the United States, are ranked as the co-first inventors, having recorded 18 patent documents each. In second place, the inventor, Law Richard Jin, from the United States, has recorded 13 patent documents.

Table 2. Inventors (top 10) of resultant patents as a function of patent documents for hydrogel-based bioinks.

Inventors	Patent Documents
Murphy Keith	18
Dorfman Scott	18
Law Richard Jin	13
Martinez Hector	10
Gatenholm Erik	9
Utama Robert Hadinoto	7
Fife Christopher Michael	7
Ribeiro Julio Cesar Caldeira	7
Atapattu Lakmali	7
Le Vivian Anne	7

All the found patent documents of these three above inventors concern the healthcare company Organovo INC (Solana Beach, CA, USA) as applicants and/or owners. It specializes in the design and development of human tissues for in vitro and therapeutic applications, such as preclinical testing and drug discovery research, by utilizing 3D bioprinting technology.

3.4. Applicants

The applicant is a person (i.e., a natural person) or an organization (i.e., a legal entity) that has filed a patent application. In several cases, the applicant can also be the inventor, and there may be more than one applicant per patent application [11].

For hydrogel-based bioinks, the top 10 applicants until 2020 are presented in Table 3. Regarding this top 10, all applicants are considered as organizations: companies, foundations, academic institutions, or universities.

Table 3. Applicants (top 10) of resultant patents as a function of patent documents for hydrogel-based bioinks.

Applicants	Patent Documents
Organovo INC	19
Cellink AB	10
Inventia Life Science PTY LTD	7
POSTECH Academy-Industry Foundation	5
Advanced Solutions Life Sciences LLC	5
Korea Institute of Science and Technology	5
Hangzhou Meizhuo Biotechnology CO LTD	3
Ulsan National Institute of Science and Technology	3
Texas A&M University System	3
Revotek CO LTD	2

The company Organovo INC (Solana Beach, CA, USA), as a legal entity, is ranked as the first applicant who has recorded 19 patent documents. In second place, the company Cellink AB (Gothenburg, Sweden), as a legal entity, has recorded 10 patent documents. Thirdly, the company Inventia Life Science PTY LTD (Alexandria, Australia), as a legal entity, has recorded seven patent documents.

3.5. Owners

The assignee, or patent owner, is a person (i.e., a natural person) or an organization (i.e., a legal entity) to whom the inventor or applicant assigned the right to a patent. The patent owner has the right, for a period limited to the duration of the patent term, to protect his brainchild. The patent system prohibits others from making, using, or selling the invention without the inventor's permission, or it requires others to use the invention under terms agreed upon with the inventor [12].

For hydrogel-based bioinks, the top 10 owners until 2020 are presented in Table 4. Regarding this top 10, all owners are considered organizations: companies, universities, foundations, or government bodies.

Table 4. Owners (top 10) of resultant patents as a function of patent documents for hydrogel-based bioinks.

Owners	Patent Documents
Organovo INC	8
Cellink AB	3
Inventia Life Science PTY LTD	2
Medical University of South Carolina	2
Advanced Solutions Life Sciences LLC	2
Texas A&M University System	2
MUSC Foundation for Research Development	2
Revotek CO LTD	2
Board of Regents of the University of Texas System	2
Curators of the University of Missouri	2

The company Organovo INC (Solana Beach, CA, USA), as a legal entity, is ranked as the first owner, having recorded eight patent documents. In second place, the company Cellink AB (Gothenburg, Sweden), as a legal entity, has recorded three patent documents. As for the podium of the third place, it is shared between eight legal entities that are: Inventia Life Science PTY LTD (Alexandria, Australia), Medical University of South Carolina (Charleston, SC, USA), Advanced Solutions Life Sciences LLC (Louisville, KY, USA), Texas A&M University System (College Station, TX, USA), MUSC Foundation for Research Development (Charleston, SC, USA), Revotek CO LTD (Lewes, DE, USA), Board of Regents of the University of Texas System (Austin, TX, USA), and Curators of the University of Missouri (Columbia, MO, USA), with two patent documents each.

3.6. Patent Classifications

The International Patent Classification (IPC) is a hierarchical system in the form of codes, which divides all technology areas into a range of sections, classes, subclasses, groups, and subgroups. It is an international classification system that provides standard information to categorize inventions and evaluate their technological uniqueness [13,14].

For hydrogel-based bioinks, the top 10 IPC codes until 2020 are presented in Table 5. The most IPC code corresponds to A61K9/52 which is a subgroup of materials for prostheses or for coating prostheses characterized by their function or physical properties. More specifically, it concerns hydrogels or hydrocolloids. This subgroup recorded, alone, 38 patent documents. The second IPC code corresponds to A61L27/38, which is a subgroup of materials for prostheses or for coating prostheses containing ingredients of undetermined constitution or reaction products thereof, such as animal cells, has recorded 34 patent documents.

Table 5. IPC codes (top 10) of resulted patents concerning hydrogel-based bioinks as a function of patent documents with the meaning of each IPC code [13].

IPC	Description	Patent Documents
A61L27/52	Materials for prostheses or for coating prostheses characterized by their function or physical properties: hydrogels or hydrocolloids.	38
A61L27/38	Materials for prostheses or for coating prostheses containing ingredients of undetermined constitution or reaction products thereof: animal cells.	34
B33Y80/00	Products made by additive manufacturing.	32
B33Y10/00	Processes of additive manufacturing.	30
B33Y70/00	Materials specially adapted for additive manufacturing.	26
A61L27/20	Materials for prostheses or for coating prostheses in the form of macromolecular materials: polysaccharides.	20
C12M1/00	Apparatus for enzymology or microbiology.	18
C12N5/00	Undifferentiated human, animal or plant cells (e.g., cell lines; tissues; cultivation or maintenance thereof; culture media therefor).	18
B33Y30/00	Apparatus for additive manufacturing; details thereof or accessories therefor.	18
C12M3/00	Tissue, human, animal or plant cell, or virus culture apparatus.	17

4. Conclusions

This patent analysis concerned only the innovation and improvement of hydrogel-based bioinks until 2020. A detailed analysis of the patentability of the used hydrogels, their preparation methods and their formulations, as well as the 3D bioprinting process using hydrogels, have been provided. During the search, 119 patent documents were found (103 patent applications and 16 granted patents). The United States was ranked first with 32 patent documents, and 2020 was the year with the maximum number of patent documents (31).

The innovation and improvement of hydrogel-based bioinks concerned raw materials (synthetic and natural polymers), synthesis and methods of preparation, as well as formulations and fabrication processes. Based on the patent classification, all filled patents and most inventions are intended for materials for prostheses or for coating prostheses, characterized firstly by their function or physical properties, such as hydrogels or hydrocolloids, and containing secondly ingredients of undetermined constitution or reaction products thereof, such as animal cells. Knowledge clusters and expert driving factors show that research based on additive manufacturing products, processes, and materials specially adapted for additive manufacturing is concentrated in the most patents.

Supplementary Materials: The following supporting information can be downloaded at: https://www.mdpi.com/article/10.3390/IOCPS2021-11239/s1.

Funding: This research received no external funding.

Institutional Review Board Statement: Not applicable.

Informed Consent Statement: Not applicable.

Data Availability Statement: The data presented in this study are available within this article content.

Acknowledgments: The author acknowledges the World Intellectual Property Organization for the Patentscope search service and the Cambia Institute for The Lens patent data set used in this study.

Conflicts of Interest: The author declares that this article content has no conflict of interest. The author has no relevant affiliations or financial involvement with any organization or entity with a financial interest in or financial conflict with the subject matter or materials discussed in this article.

References

1. Zhang, S.; Huang, D.; Lin, H.; Xiao, Y.; Zhang, X. Cellulose Nanocrystal Reinforced Collagen-Based Nanocomposite Hydrogel with Self-Healing and Stress-Relaxation Properties for Cell Delivery. *Biomacromolecules* **2020**, *21*, 2400–2408. [CrossRef] [PubMed]
2. Roehm, K.D.; Madihally, S.V. Bioprinted chitosan-gelatin thermosensitive hydrogels using an inexpensive 3D printer. *Biofabrication* **2017**, *10*, 015002. [CrossRef] [PubMed]
3. Narayanan, L.K.; Huebner, P.; Fisher, M.B.; Spang, J.T.; Starly, B.; Shirwaiker, R.A. 3D-Bioprinting of Polylactic Acid (PLA) Nanofiber–Alginate Hydrogel Bioink Containing Human Adipose-Derived Stem Cells. *ACS Biomater. Sci. Eng.* **2016**, *2*, 1732–1742. [CrossRef] [PubMed]
4. Skardal, A.; Zhang, J.; Prestwich, G.D. Bioprinting vessel-like constructs using hyaluronan hydrogels crosslinked with tetrahedral polyethylene glycol tetracrylates. *Biomaterials* **2010**, *31*, 6173–6181. [CrossRef] [PubMed]
5. Lee, C.; Abelseth, E.; de la Vega, L.; Willerth, S.M. Bioprinting a novel glioblastoma tumor model using a fibrin-based bioink for drug screening. *Mater. Today Chem.* **2019**, *12*, 78–84. [CrossRef]
6. Lee, V.; Singh, G.; Trasatti, J.P.; Bjornsson, C.; Xu, X.; Tran, T.N.; Yoo, S.-S.; Dai, G.; Karande, P. Design and Fabrication of Human Skin by Three-Dimensional Bioprinting. *Tissue Eng. Part C Methods* **2013**, *20*, 473–484. [CrossRef] [PubMed]
7. Forgacs, G.; Jakab, K.; Neagu, A.; Mironov, V. Self-assembling Cell Aggregates and Methods of Making Engineered Tissue Using the Same. Granted Patent U.S. 8241905 B2, 14 August 2012.
8. World Intellectual Property Organization. Patentscope. Available online: https://patentscope.wipo.int (accessed on 2 September 2021).
9. World Intellectual Property Organization. Patentscope Fields Definition. Available online: https://patentscope.wipo.int/search/en/help/fieldsHelp.jsf (accessed on 2 September 2021).
10. Cambia Institute. The Lens Patent Data Set. Version 8.0.14. Available online: https://www.lens.org (accessed on 2 September 2021).
11. European Patent Office. Espacenet Glossary. Version 1.24.1. Available online: https://worldwide.espacenet.com/patent (accessed on 2 September 2021).
12. World Intellectual Property Organization. What Is Intellectual Property? Frequently Asked Questions: Patents. Available online: https://www.wipo.int/patents/en/faq_patents.html (accessed on 2 September 2021).
13. World Intellectual Property Organization. IPC Publication. IPCPUB v8.5. Available online: https://www.wipo.int/classifications/ipc/ipcpub (accessed on 2 September 2021).
14. World Intellectual Property Organization. Guide to the International Patent Classification (IPC). Available online: https://www.wipo.int/edocs/pubdocs/en/wipo_guide_ipc_2020.pdf (accessed on 2 September 2021).

materials proceedings

MDPI

Proceeding Paper

RAFT-Mediated Radiation Grafting on Natural Fibers in Aqueous Emulsion [†]

Bin Jeremiah D. Barba [1,2,*], David P. Peñaloza, Jr. [2], Noriaki Seko [3] and Jordan F. Madrid [1]

[1] Chemistry Research Section, Department of Science and Technology, Philippine Nuclear Research Institute, Quezon City 1101, Philippines; jfmadrid@pnri.dost.gov.ph

[2] Department of Chemistry, College of Science, De La Salle University Manila, Manila 0922, Philippines; david.penaloza.jr@dlsu.edu.ph

[3] Quantum Beam Science Research Directorate, National Institutes for Quantum and Radiological Science and Technology (QST), Takasaki 370-1292, Japan; seko.noriaki@qst.go.jp

* Correspondence: bjdbarba@pnri.dost.gov.ph; Tel.: +63-289-296-011

† Presented at the 2nd International Online Conference on Polymer Science—Polymers and Nanotechnology for Industry 4.0, 1–15 November 2021; Available online: https://iocps2021.sciforum.net/.

Abstract: Using aqueous emulsion as the medium in radiation-induced graft polymerization (RIGP) offers an environment-friendly shift from organic solvents while increasing polymerization efficiency through known water radiolysis-based graft initiation. In the present paper, we further extend the applicability of RIGP in emulsion under the influence of reversible addition fragmentation chain transfer (RAFT) mechanisms. Emulsions prepared with Tween 20 showed good colloidal stability for several hours. Subjecting it to simultaneous irradiation with abaca fibers resulted in successful grafting, supported by gravimetric, IR, SEM, and TG analysis. A correlation was drawn between smaller monomer micelles and the enhancement of grafting driven by diffusion and surface area coverage. RAFT mechanisms were also conserved based on molecular weight evolution. RAFT-mediated RIGP in aqueous emulsion shows good potential as a versatile and green surface modification technique for natural fibers for various functional applications.

Keywords: radiation-induced graft polymerization (RIGP); reversible addition fragmentation chain transfer (RAFT) polymerization; emulsion grafting; natural fibers

Citation: Barba, B.J.D.; Peñaloza, D.P., Jr.; Seko, N.; Madrid, J.F. RAFT-Mediated Radiation Grafting on Natural Fibers in Aqueous Emulsion. *Mater. Proc.* **2021**, *7*, 4. https://doi.org/10.3390/IOCPS2021-11243

Academic Editor: Shin-ichi Yusa

Published: 30 October 2021

Publisher's Note: MDPI stays neutral with regard to jurisdictional claims in published maps and institutional affiliations.

1. Introduction

Emulsion polymerization involves the free radical polymerization of monomer molecules in large and discrete polymer particles dispersed in a continuous aqueous phase stabilized by surfactants. This aids in the utilization of water as the solvent for relatively hydrophobic monomers contributing to green chemistry, as it significantly reduces organic solvent dependence and their corresponding VOC generation [1]. Due to the segregation of free radicals within the micelles formed in emulsions, the probability of bimolecular termination is reduced, leading to faster polymerization rates and higher molecular weights [2]. This technique has been successfully applied in various industries utilizing polymers that include coatings, adhesives, plastics, and synthetic rubber.

This has also been explored in graft copolymerization, particularly radiation-induced graft polymerization (RIGP). RIGP can be applied in the synthesis of functional materials as it involves the attachment of polymer chains with advantageous tailored properties to the surface of a base polymer with desirable bulk character. The general mechanism is illustrated in Figure 1 [3]. The key advantage here is that graft reactivity is usually higher in water due to its radiolytic products, which participate in the initiation phase [4]. Glycidyl methacrylate (GMA), a widely used monomer in RIGP, is only soluble in water at concentrations less than 2%. In a previous work [5], emulsifying GMA with Tween 20 formed a milky emulsion that was stable for 48 h and resulted in a sufficiently high grafting yield on polyethylene fibers at doses as low as 10 kGy. GMA concentrations can reach up to 10% (w/w), which effectively enhances the degree of grafting [6,7].

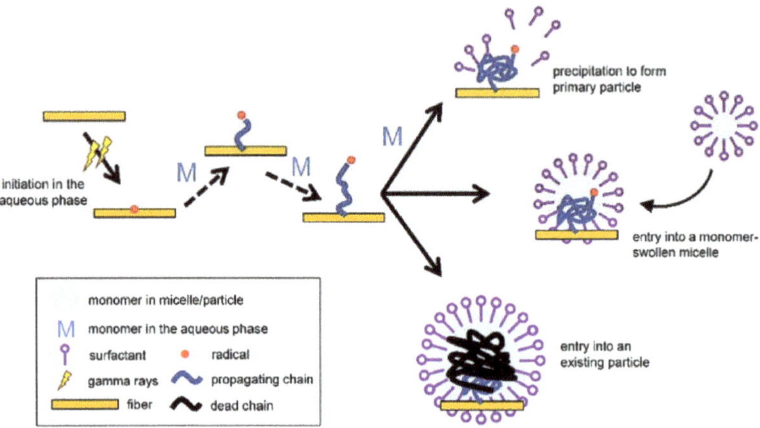

Figure 1. Proposed general mechanism of RIGP in emulsion [3].

In the preparation of diverse functional materials, control over polymer brush architecture leads to a significant influence over surface properties and often enhances its specific functionality [8]. Among controlled radical polymerization techniques, the reversible addition fragmentation chain transfer (RAFT) mechanism is applicable involving both radiation-initiated reactions and emulsion systems [9–11]. The RAFT agent, initially dissolved in the monomer before emulsification, is likely to be found inside the micelle, where it can mediate chain growth through a reversible deactivation of propagating chains [12]. Difficulties associated with the use of RAFT in ab initio emulsion systems, such as low polymerization rates, broad molecular weight distribution, and loss of control, are usually attributed to colloidal instability and RAFT transport [13]. In this work, RAFT-mediated radiation grafting in aqueous emulsion using Tween 20 as the surfactant for emulsion polymerization was assessed for its pseudo-living character and its potential as a green surface modification technique for natural fibers for various functional applications.

2. Materials and Methods

2.1. Emulsion Preparation

The emulsion was prepared with the following components by weight: 10% GMA (>95%, TCI, Tokyo, Japan), 0.2% 4-cyano-4-(carbonothioylthio)pentanoic acid (CPPA, 97%, Strem Chemicals Inc., Newburyport, MA, USA), and 0.5–2% Tween 20 (Kanto Chemical Co., Inc., Tokyo, Japan) in ultrapure water. The solution was homogenized at 6000 rpm for 10 min. Stability and micelle size were measured by dynamic light scattering using an FPAR 1000 particle size analyzer (Otsuka Electronics Co., Ltd., Osaka, Japan) at 25 °C.

2.2. Radiation Grafting of Abaca Fibers

Pre-weighed abaca woven fabrics (1×5 cm^2) and 8 g of the emulsion were placed in a glass vial and purged with N$_2$ gas. The vial was placed in a Co-60 gamma irradiation chamber and polymerization was carried out at a dose rate of 1 kGy/h (in air). The fabrics were washed several times with methanol and tetrahydrofuran (THF) and dried to constant weight, in vacuo. The homopolymers were then precipitated from the solution using methanol and weighed. The degree of grafting (DG, %) was calculated from the percent weight gain of the fabric, while conversion (%) was determined from the added weight of the collected homopolymer and weight gain of the fabric over the initial amount of monomer in the vial. A sample of homopolymers was dissolved in THF and analyzed using gel permeation chromatography (Chromaster, Hitachi HighTech, Japan equipped with Shodex Asahipak GF-16 7B guard and two GF-7M columns). Polymethyl methacrylate standards were used.

2.3. Characterization

The abaca fabrics before and after RIGP were subjected to several characterization techniques. The chemical information of the fabrics was examined by a Frontier FTIR Spectrophotometer (Perkin Elmer, Japan) in attenuated total reflectance (ATR) mode with 4 cm^{-1} resolution. The morphology and elemental composition of the fabric surface were examined by scanning electron microscopy with an energy dispersive X-ray (SEM-EDX) taken by a Hitachi SU3500 scanning electron microscope (Hitachi HighTech, Tokyo, Japan), coupled with an X-Max EDX spectrometer (Horiba Ltd., Kyoto, Japan) at vacuum conditions and acceleration voltage of 50 Pa and 10 kV, respectively. All fabrics were gold coated using a JFC-1600 Auto Fine Coater (JEOL Ltd., Tokyo, Japan) prior to scanning. The thermal decomposition profile of the fabrics was examined using a TG/DTA 6200 Extar 6000 (Seiko Instruments Inc., Tokyo, Japan) under a nitrogen atmosphere using a 5 mm aluminum sample pan, with the following heating program: (1) 25–130 °C at 20 °C/min; (2) 130 °C for 10 min; and continuing to 550 °C at 10 °C/min.

3. Results and Discussion

In the mechanism of emulsion polymerization via RAFT, the formation of oligomers at the initial phase can dramatically reduce the chemical potential of the nucleated particle and lead to a large amount of monomer to transfer from droplets to these particles [14]. This phenomenon, known as superswelling, ultimately destabilizes the emulsion and often leads to a loss of control of polymerization associated with the formation of a colored layer or coagulum. The use of a relatively higher surfactant concentration, preferably nonionic like Tween 20 above its critical micelle concentration (CMC), has been known to circumvent superswelling [15]. High amounts of surfactant (at least greater than their CMC) also ensure that most of the monomers, as well as the RAFT agent are in the stable micelles (<1 μm), which are considerably smaller than monomer droplets (>1 μm). This also aids in the rapid transport of RAFT agent to the growing particles owing to the larger surface area of the micelles [11].

Emulsion systems with different surfactant-to-monomer ratios were observed. As seen in Figure 2a, the degree of grafting (DG) increased when the surfactant amount was changed from 0.5 to 1% and then proceeded to decrease at higher levels of Tween 20. DG seems to correlate inversely with the observed micelle size of the emulsion. Smaller micelles can cover a wider surface area of the trunk polymer leading to an enhanced grafting efficiency [5]. Furthermore, smaller micelles usually diffuse better throughout the solution allowing them greater access to the site of graft propagation [4,5,16]. Based on these results, the surfactant concentration was fixed to 1% in further experiments as it exhibited the highest DG. At this concentration, the emulsion remained in a milky state (Figure 2b) with no phase separation, and was relatively stable for at least 10 h.

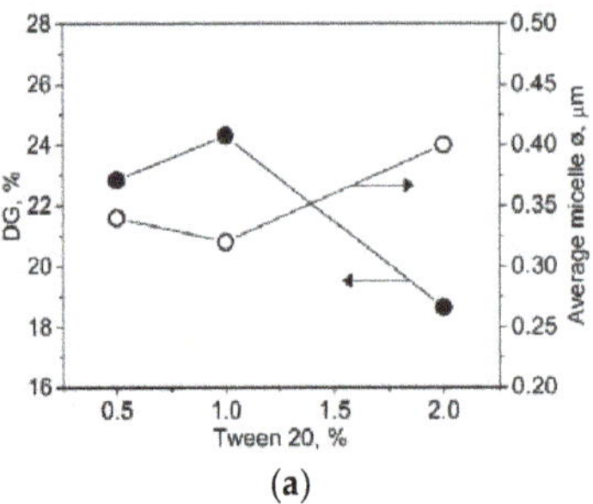

(a)   (b)

Figure 2. CTPA/GMA emulsion studies: (**a**) effect of surfactant concentration on the degree of grafting (DG) at 5 kGy absorbed dose and average micelle diameter ∅; and (**b**) representative images of solution during initial separated phases and emulsified milky state.

Grafting in emulsion was further carried out at various absorbed doses. As seen in Figure 3a, there is an initial rapid increase in DG with the absorbed dose as it correlates to radical production. Continuous chain growth is also a feature of RAFT-mediated polymerization, which minimizes irreversible chain termination (dead chains) through the reversible deactivation of propagating polymer radicals ("living" chains) [17,18]. Therefore, grafted polymer chains continue to grow and contribute to the DG as long as monomers remain available. At higher absorbed doses, DG increase slowed down possibly due to the decrease in available monomer molecules during the course of the polymerization. This is reflected in the monomer conversion values shown in Figure 2b. Similarly, monomer conversion showed a rapid increase in the absorbed dose, then slowed down at higher doses due to restricted monomer diffusion with an increase in polymer concentration and viscosity. There was an initial linear increase in the conversion values as a function of absorbed dose, which translates to polymerization time (h) as the dose rate is fixed to 1 kGy/h. To estimate the properties of the grafted polymers in terms of molecular weight and polydispersity, GPC analysis was performed on the precipitated homopolymers from the solution. As reported in previous works on RAFT-mediated radiation-induced grafting, the growth of surface grafted polymer chains exists in dynamic equilibrium with the free polymer chains (homopolymers) formed in the grafting solution [9,19]. Figure 3c shows the chromatograms of PGMA homopolymers formed using conventional (without CPPA) and RAFT-mediated RIGP. In contrast to the former's broad and multimodal molecular weight distribution, using RAFT polymerization demonstrated narrow and monomodal distribution with a polydispersity index of 1.35, and good correspondence with the calculated value based on the RAFT-monomer ratio and conversion [12]. These results suggests the pseudo-living characteristics of the polymerization, indicating that the RAFT mechanism was in effect in the emulsion [20].

Analyzing the fabrics using FTIR (Figure 4a), successful grafting is evident as supported by the appearance of C=O stretching bands at 1725 cm^{-1} and epoxide ring deformations at 750–950 cm^{-1}, in addition to the cellulose profile of the fiber shown as peaks at 1600, 1110, and 1040 cm^{-1}, attributed to the glucose ring vibrations [21]. Figure 4b,c also show the extension of the degradation profile of abaca fiber after the cellulose main chain 250–360 °C to 420 °C attributed to the degradation of the grafted polymers [22]. Finally, the abaca fibers were examined using SEM-EDX. The pristine fiber showed a relatively smooth surface made of 60% carbon and 40% oxygen atoms (Figure 4d. Meanwhile, abaca-*g*-PGMA showed a much rougher surface due to the amorphous PGMA covalently grafted on its surface (Figure 4e). There is also an increase in the carbon content from the PGMA chain, shifting the composition to 70% C and 30% O. Trace amounts of sulfur (0.02%) were also detected which can be attributed to the thiocarbonylthio end of CPPA, which adds to the propagating chain during RAFT-mediated polymerization [9,12,18].

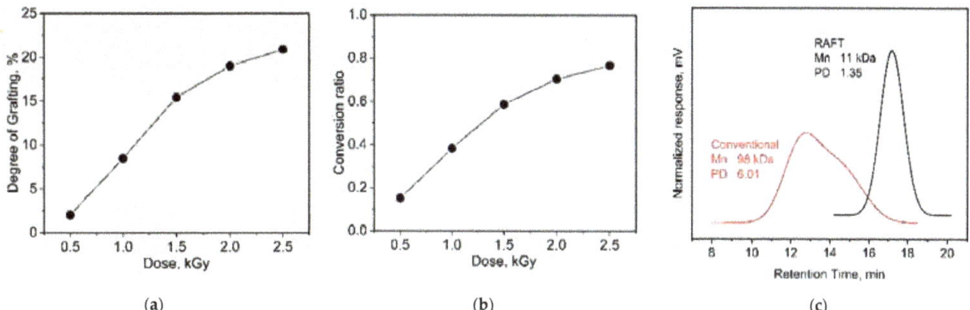

Figure 3. Effect of absorbed dose on the (**a**) degree of grafting and (**b**) monomer conversion; and (**c**) representative GPC chromatograms for PGMA formed using the conventional technique and with CPPA at ~0.8 conversion.

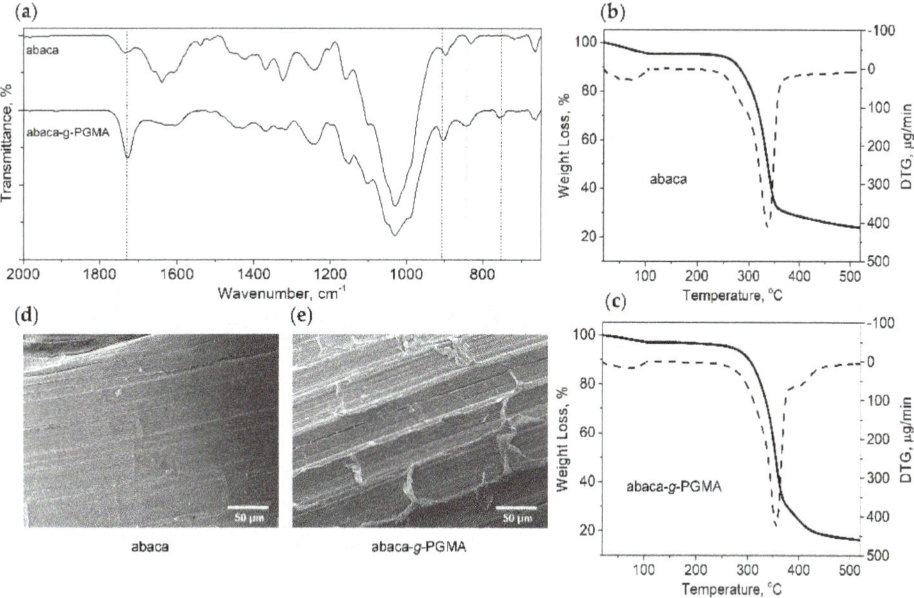

Figure 4. Evidence of grafting: (**a**) FTIR spectra highlighting the appearance of carbonyl and epoxide peaks on the (dashed lines); (**b,c**) weight loss and DTG (dashed lines) profiles; and (**d,e**) SEM micrograph of fiber surface of pristine and grafted abaca (DG, ~20%).

4. Conclusions

Graft polymerization reactions performed in aqueous emulsion offer an environment-friendly alternative to the use of organic solvents. In this work, glycidyl methacrylate was successfully grafted from abaca fibers using radiation-induced graft polymerization, mediated by reversible addition fragmentation chain transfer mechanism. Stable milky emulsions were formed using Tween 20 as the surfactant, and CPPA as the chain transfer agent. The emulsion remained stable for several hours and did not form an oily layer or coagulum indicative of unsuccessful RAFT polymerization. Instead, grafting proceeded with pseudo-living characteristics as supported by the increase in the degree of grafting with the linear increase in conversion, and the narrow and monomodal molecular weight distribution of generated homopolymers with good correspondence to the theoretic value.

Evidence of grafting was verified using FTIR, TGA, and SEM-EDX, which all showed the successful attachment of PGMA chains onto the abaca fiber. These results highlight the potential of RAFT-mediated RIGP in aqueous emulsion as a green surface modification technique for natural fibers. Studies on the further functionalization and characterization of the fibers for applications as adsorbents and composite reinforcement fillers are currently underway.

Author Contributions: Conceptualization, B.J.D.B., J.F.M. and D.P.P.J.; methodology, B.J.D.B., J.F.M. and N.S.; formal analysis, B.J.D.B.; investigation, B.J.D.B.; resources, N.S.; writing—original draft preparation, B.J.D.B.; writing—review and editing, J.F.M., D.P.P.J. and N.S.; supervision, J.F.M. and D.P.P.J. All authors have read and agreed to the published version of the manuscript.

Institutional Review Board Statement: Not applicable.

Informed Consent Statement: Not applicable.

Data Availability Statement: The data presented in this study are available on request from the corresponding author.

Acknowledgments: The authors acknowledge the financial support given by the Department of Science and Technology—Science Education Institute (DOST-SEI), which sponsored the graduate scholarship and fellowship grant of Barba as part of their Human Resource Development Program. This work was also partially supported by JSPS KAKENHI (Grant No. JP17K05956). The authors also acknowledge the contributions of the Environment Polymer Group of QST-TARRI in providing assistance and facilities for the conduct of experiments.

Conflicts of Interest: The authors declare no conflict of interest.

References

1. Zhang, Y.; Dubé, M.A. *Green Emulsion Polymerization Technology BT—Polymer Reaction Engineering of Dispersed Systems: Volume I*; Pauer, W., Ed.; Springer International Publishing: Cham, Switzerland, 2018; pp. 65–100, ISBN 978-3-319-73479-8.
2. Chern, C.S. Emulsion polymerization mechanisms and kinetics. *Prog. Polym. Sci.* **2006**, *31*, 443–486. [CrossRef]
3. Lovell, P.A.; Schork, F.J. Fundamentals of Emulsion Polymerization. *Biomacromolecules* **2020**, *21*, 4396–4441. [CrossRef] [PubMed]
4. Wada, Y.; Tamada, M.; Seko, N.; Mitomo, H. Emulsion grafting of vinyl acetate onto preirradiated poly(3-hydroxybutyrate) film. *J. Appl. Polym. Sci.* **2008**, *107*, 2289–2294. [CrossRef]
5. Seko, N.; Bang, L.T.; Tamada, M. Syntheses of amine-type adsorbents with emulsion graft polymerization of glycidyl methacrylate. *Nucl. Instrum. Methods Phys. Res. Sect. B Beam Interact. Mater. Atoms* **2007**, *265*, 146–149. [CrossRef]
6. Madrid, J.F.; Lopez, G.E.P.; Abad, L.V. Application of full-factorial design in the synthesis of polypropylene-g-poly(glycidyl methacrylate) functional material for metal ion adsorption. *Radiat. Phys. Chem.* **2017**, *136*, 54–63. [CrossRef]
7. Madrid, J.F.; Ueki, Y.; Seko, N. Abaca/polyester nonwoven fabric functionalization for metal ion adsorbent synthesis via electron beam-induced emulsion grafting. *Radiat. Phys. Chem.* **2013**, *90*, 104–110. [CrossRef]
8. Feng, C.; Huang, X. Polymer Brushes: Efficient Synthesis and Applications. *Acc. Chem. Res.* **2018**, *51*, 2314–2323. [CrossRef] [PubMed]
9. Barsbay, M.; Güven, O.; Stenzel, M.H.; Davis, T.P.; Barner-Kowollik, C.; Barner, L. Verification of controlled grafting of styrene from cellulose via radiation-induced RAFT polymerization. *Macromolecules* **2007**, *40*, 7140–7147. [CrossRef]
10. Madrid, J.F.; Ueki, Y.; Abad, L.V.; Yamanobe, T.; Seko, N. RAFT-mediated graft polymerization of glycidyl methacrylate in emulsion from polyethylene/polypropylene initiated with γ-radiation. *J. Appl. Polym. Sci.* **2017**, *134*, 45270. [CrossRef]
11. Urbani, C.N.; Nguyen, H.N.; Monteiro, M.J. RAFT-Mediated Emulsion Polymerization of Styrene using a Non-Ionic Surfactant. *Aust. J. Chem.* **2006**, *59*, 728. [CrossRef]
12. Perrier, S. 50th Anniversary Perspective: RAFT Polymerization—A User Guide. *Macromolecules* **2017**, *50*, 7433–7447. [CrossRef]
13. Prescott, S.W.; Ballard, M.J.; Rizzardo, E.; Gilbert, R.G. RAFT in Emulsion Polymerization: What Makes it Different? *Aust. J. Chem.* **2002**, *55*, 415–424. [CrossRef]
14. Luo, Y.; Tsavalas, J.; Schork, F.J. Theoretical aspects of particle swelling in living free radical miniemulsion polymerization. *Macromolecules* **2001**, *34*, 5501–5507. [CrossRef]
15. Luo, Y.; Cui, X. Reversible addition–fragmentation chain transfer polymerization of methyl methacrylate in emulsion. *J. Polym. Sci. Part A Polym. Chem.* **2006**, *44*, 2837–2847. [CrossRef]
16. Ma, H.; Morita, K.; Hoshina, H.; Seko, N. Synthesis of Amine-type Adsorbents with Emulsion Graft Polymerization of 4-hydroxybutyl Acrylate Glycidylether. *Mater. Sci. Appl.* **2011**, *2*, 776–784. [CrossRef]
17. Moad, G.; Barner-Kowollik, C. The Mechanism and Kinetics of the RAFT Process: Overview, Rates, Stabilities, Side Reactions, Product Spectrum and Outstanding Challenges. In *Handbook of RAFT Polymerization*; John Wiley & Sons, Inc.: Hoboken, NJ, USA, 2008; pp. 51–104.

18.   Barner, L.; Zwaneveld, N.; Perera, S.; Pham, Y.; Davis, T.P. Reversible addition–fragmentation chain-transfer graft polymerization of styrene: Solid phases for organic and peptide synthesis. *J. Polym. Sci. Part A Polym. Chem.* **2002**, *40*, 4180–4192. [CrossRef]
19.   Li, Y.; Schadler, L.S.; Benicewicz, B.C. Surface and particle modification via the RAFT process: Approach and properties. In *Handbook of RAFT Polymerization*; Wiley Online Library: Hoboken, NJ, USA, 2008.
20.   Madrid, J.F.; Barsbay, M.; Abad, L.; Güven, O. Grafting of N, N-dimethylaminoethyl methacrylate from PE/PP nonwoven fabric via radiation-induced RAFT polymerization and quaternization of the grafts. *Radiat. Phys. Chem.* **2016**, *124*, 145–154. [CrossRef]
21.   Sharif, J.; Mohamad, S.F.; Fatimah Othman, N.A.; Bakaruddin, N.A.; Osman, H.N.; Güven, O. Graft copolymerization of glycidyl methacrylate onto delignified kenaf fibers through pre-irradiation technique. *Radiat. Phys. Chem.* **2013**, *91*, 125–131. [CrossRef]
22.   Ananthalakshmi, N.R.; Wadgaonkar, P.P.; Sivaram, S.; Varma, I.K. Thermal behaviour of glycidyl methacrylate homopolymers and copolymers. *J. Therm. Anal. Calorim.* **1999**, *58*, 533–539. [CrossRef]

materials proceedings

Proceeding Paper

Nanografting of Polymer Brushes on Gold Substrate by RAFT-RIGP [†]

Bin Jeremiah D. Barba [1,*], Patricia Nyn L. Heruela [2], Patrick Jay E. Cabalar [1], John Andrew A. Luna [1], Allan Christopher C. Yago [2] and Jordan F. Madrid [1]

[1] Chemistry Research Section, Philippine Nuclear Research Institute, Department of Science and Technology, Quezon City 1101, Philippines; pjecabalar@pnri.dost.gov.ph (P.J.E.C.); jluna@pnri.dost.gov.ph (J.A.A.L.); jfmadrid@pnri.dost.gov.ph (J.F.M.)

[2] Institute of Chemistry, College of Science, University of the Philippines Diliman, Quezon City 1101, Philippines; plheruela@up.edu.ph (P.N.L.H.); acyago@up.edu.ph (A.C.C.Y.)

* Correspondence: bjdbarba@pnri.dost.gov.ph; Tel.: +63-2-8929-6011

[†] Presented at the 2nd International Online Conference on Polymer Science—Polymers and Nanotechnology for Industry 4.0, 1–15 November 2021; Available online: https://iocps2021.sciforum.net/.

Citation: Barba, B.J.D.; Heruela, P.N.L.; Cabalar, P.J.E.; Luna, J.A.A.; Yago, A.C.C.; Madrid, J.F. Nanografting of Polymer Brushes on Gold Substrate by RAFT-RIGP. *Mater. Proc.* **2021**, 7, 5. https://doi.org/10.3390/IOCPS2021-11587

Academic Editor: Shin-ichi Yusa

Published: 5 November 2021

Abstract: Optical sensors based on surface plasmon resonance (SPR) have made great strides in the detection of various chemical and biological analytes. A surface plasmon is a bound, non-radiative evanescent wave generated as resonant electrons on a metal–dielectric surface to absorb energy from an incident light. As analytes bind to a functionalized metal substrate, the refractometric response generated can be used for quantitation with great selectivity, sensitivity, and capacity for label-free real-time analysis. Polymer nanobrushes are ideal recognition elements because of their greater surface area and their wide range of functional versatility. Here, we introduce a simple "grafting-from" method to covalently attach nanometer-thick polymer chains on a gold surface. Nanografting on gold-coated BK-7 glass was performed in two steps: (1) self-assembly of organosulfur compounds; and (2) RAFT-mediated radiation-induced graft polymerization (RAFT-RIGP) of polyglycidyl methacrylate (PGMA). Surface modification was monitored and verified using FTIR and SPR. Layer-by-layer thickness calculated based on Winspall 3.02 simulation fitted with experimental SPR curves showed successful self-assembly of 1-dodecanethiol (DDT) monolayer with thickness measuring 1.4 nm. These alkane chains of DDT served as the graft initiation sites for RAFT-RIGP. Nanografting was controlled by adjusting the absorbed dose in the presence of chain transfer agent, 4-cyano-4-(phenylcarbonothioylthio)pentanoic acid. The molecular weight of grafted polymers measuring 2.8 and 4.3 kDa corresponded to a thickness increase of 3.6 and 7.9 nm, respectively. These stable nanografted gold substrates may be further functionalized for sensing applications.

Keywords: surface plasmon resonance; nanografting; radiation-induced graft polymerization; reversible addition fragmentation chain transfer polymerization; optical sensors

1. Introduction

Surface plasmon resonance (SPR) sensor technology for the detection of chemical and biological substances has been receiving growing interest from the scientific community. It has been increasingly employed, not only in gas sensing, but also in many other important applications in food safety, biology, and medical diagnostics [1]. The overall performance of an SPR sensor is affected by the characteristics of the surface functionalization. The most commonly used approach is through covalent attachment of the receptor molecules/functional groups on the metallic film, supporting the propagation of charge density waves [2]. Covalent immobilization strategies include chemical reactions such as amine, aldehyde or thiol coupling on previously formed functional self-assembled monolayers on gold surfaces. The use of polymers to modify the gold layer is also an alternative.

This can be carried out through surface-initiated polymerization or a grafting-from approach based on the immobilization or production of initiator on the surface of the sensor substrate, which can be accomplished chemically or through the use of ionizing radiation. Radiation processing is a unique tool for the synthesis of nanosized polymers of various types and offers a promising alternative to conventional synthetic methods because (a) it is environmentally friendly due to the fact that it consumes less energy, (b) creates less waste, and (c) uses a lesser amount of chemicals.

Ionizing radiation can be used to initiate polymerization from solid surfaces in a process known as radiation-induced graft polymerization (RIGP). RIGP generally proceeds via a free radical polymerization process. The challenge in this process is the occurrence of irreversible termination of propagating polymer chains and chain transfer reactions that can result in a loss of control over chain length (i.e., thickness of polymer film) and chain structure together with broadening of the molecular weight distribution [3–5]. These drawbacks have led to the research area of controlled free-radical polymerization (CRP), which can be utilized to control the polymer architecture and to reduce polydispersity. One such type of CRP is reversible-addition fragmentation chain transfer (RAFT), which has been found to be compatible with ionizing radiation as the mode of initiation. In the work of Barsbay et al. [6], they demonstrated that RAFT-mediated radiation induced grafting (RAFT-RIGP) of poly(acrylic acid) (PAA) into the nanochannel walls of track-etched poly(vinylidene difluoride) (PVDF) membranes, achieving good chain length control and corresponding graft polymer nano-thickness. These features can be easily manipulated by adjusting the ratio of added RAFT agent to monomer and monomer conversion via absorbed dose [7–10]. In light of these findings, RAFT-RIGP was identified as a viable synthesis technique in the development of SPR sensors with well-defined surface characteristics. This will involve a two-step process as shown in Figure 1: (a) formation of a self-assembled layer; and (b) RAFT-RIGP from the SAM layer.

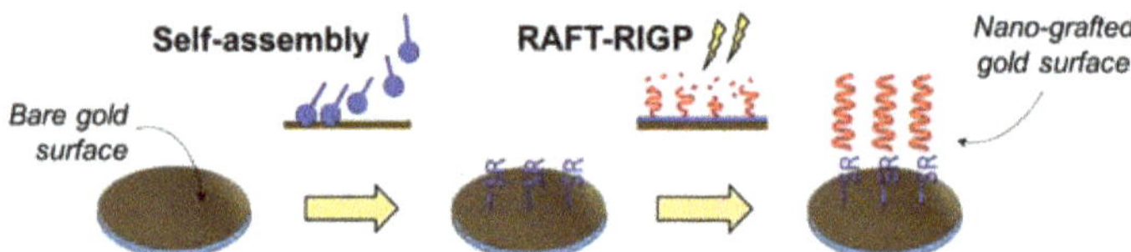

Figure 1. General synthesis scheme for the surface functionalization of gold surfaces.

2. Materials and Methods

2.1. Formation of Self-Assembled Monolayer (SAM)

A bare BK-7 gold-coated glass disc from KEI International was immersed in a solution containing 2 mM 1-dodecanethiol (DDT, Sigma Aldrich, St. Louis, MO, USA) in absolute ethanol (Ajax Finechem, Taren Point, Australia) for 5 days at room temperature with mild stirring. The disc was washed with ethanol. Drying was carried out using nitrogen gas followed by an overnight drying under vacuum.

2.2. RAFT-RIGP Adiation Grafting of Abaca Fibers

Prepared gold-SAM samples were immersed in a solution containing 3% glycidyl methacrylate (GMA, Sigma Aldrich), 0.6% Tween 20 (Scharlab) and 0.26% 4-cyano-4-(phenylcarbonothioylthio)pentanoic acid (CPPA, Sigma Aldrich) in deionized water, and emulsified at 6000 rpm for 30 min. The emulsion was bubbled with nitrogen gas for 1 h, sealed and irradiated with gamma rays at 1 kGy/h. The disc was washed with tetrahydrofuran and methanol and dried under vacuum. Homopolymers that formed in the solution were precipitated with methanol, collected, and dried.

2.3. Characterization

The gold discs were subjected to several characterization techniques after each modification step. Successful modification was verified using a Frontier FTIR spectrophotometer (Perkin Elmer, Tokyo, Japan) in ATR mode, scanning at 650–4000 cm^{-1} with 4 cm^{-1} resolution. SPR measurements were carried out using the KEI ESPRIT scanning angle SPR. Measurements were taken using 0.6% Tween 20 in deionized water at 670 nm, with the range of 4000 mdeg and angle resolution of <0.02. The resulting data were normalized using Origin® 2016. The relative resonance angle and standard deviation were taken from 2–3 measurement replicates.

2.4. Simulation and Fitting of SPR Data

Layer-by-layer thickness simulation was calculated by fitting these experimental SPR curves with a Fresnel equation algorithm (Winspall software version 3.20) [11]. Approximations were drawn from the following inputs (Table 1):

Table 1. Thickness and reflectivity constants used in Winspall Simulation.

Layer	Thickness (nm)	Refractive Index
BK-7 prism	0	1.5151 [12]
Gold	50	0.1377 [13]
DDT	0–4.0	1.4590 [14]
PGMA	0–10	1.4490 [15]
Water	0	1.3334 [12]

It should be noted that the program assumes each modification as separate layers with distinct optical constants that are not affected by the nature of their attachment. Linear regressions taken from the resonance angle and thickness of the simulated plots were used to calculate the thickness of the layers as a function of resonance angle shifts from SPR measurement. For simplicity and given limited literature values for the optical properties, pure water was considered the medium in lieu of 0.6% Tween 20 in water.

3. Results and Discussion

Self-assembled monolayers (SAMs) of thiol compounds on gold substrates are one of the most popular model systems for probing self-assembly of organic molecules on metal surfaces. The robustness of Au–S interaction between thiols and gold surfaces serves as the platform for fabrication of SAMs for diverse applications [16]. In this study, SAM serves as the base for the generation of carbon radicals through gamma radiolytic reactions. DDT is an alkanethiol with 12 C–H chains at any point of which can generate a carbon-centered radical after irradiation. Figure 2 confirms the successful SAM formation of DDT onto the gold surface as indicated by peaks at 1465 and 880 cm^{-1} corresponding to CH$_2$ bending of the alkane chain and C–S stretch of the thiol group [17]. Expected C–H stretching at around 3000 cm^{-1} was obscured by surface impurities of the bare gold specimen (not shown).

After DDT assembly on the gold surface, RAFT-RIGP was performed in simultaneous mode where the substrate is irradiated with the monomer solution such that radicals are formed on the SAM layer or monomer molecules by either direct bond breaking or reaction with solvent radiolytic species. Propagating radicals in this system may now be either free (initiated from an activated monomer molecule) or tethered (initiated from the activated substrate). Both types of chain add to the chain transfer agent, CPPA, which controls the reversible chain inactivation allowing for even growth of chains and a reduction in premature termination reactions [10,18]. FTIR analysis (Figure 2) also confirms successful grafting of PGMA onto the gold–SAM substrate with the appearance of peaks at 1730 and 1160 cm^{-1} corresponding to the C=O stretching and C–O–C vibrations of PGMA, respectively [19]. The effect of dose on grafting is shown as an increase in intensity of carbonyl peak suggesting a higher degree of grafting. This may be attributed to the progression of polymerization sustained by the RAFT mechanism, leading to an increase in molecular weight of the grafted polymer [10]. To verify, molecular weights of grafted chains were indirectly measured using the homopolymers co-generated in the reaction. As the formation of grafted polymers and homopolymers exist in dynamic equilibrium, the RAFT control exerted is expected to result in similar polymerization rates [9]. The homopolymers were analyzed using ^{1}H-NMR analysis (Figure 3), which showed the characteristic profile of PGMA as well as chemical shifts corresponding to the ring of the thiocarbonylthio end group at 7.2–8.0 ppm. The ratio between the peak area for the aromatic protons and the main chain protons was used to calculate the molecular weight of the polymers. As can be seen in Table 2, calculated values closely corresponded to theoretical values. These observations indicate the successful addition of propagating PGMA to CPPA and support that the RAFT mechanism was in effect [20].

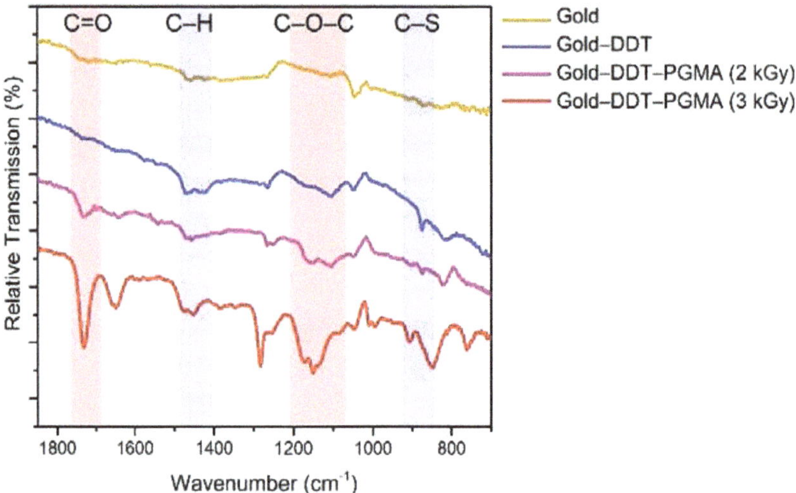

Figure 2. FTIR spectra of the gold surface before and after modification.

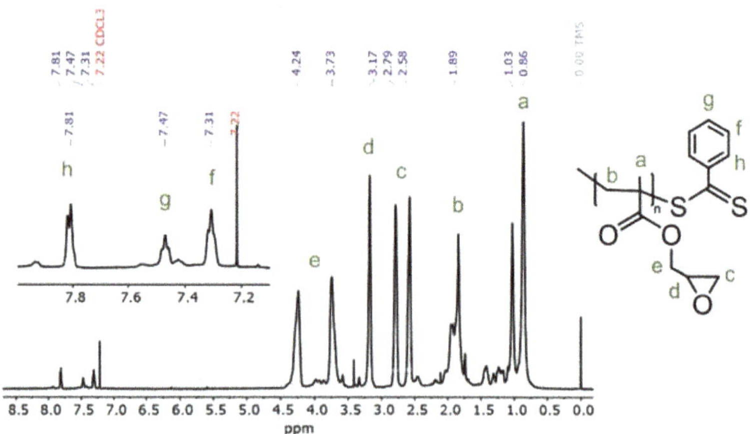

Figure 3. Representative ^{1}H-NMR profile of PGMA homopolymer generated simultaneously in the grafting solution (3 kGy).

Table 2. Effect of absorbed dose on conversion and molecular weight.

Dose, kGy	Conversion, %	Molecular Weight, kDa	
		Calculated *	NMR
2	62	2.3	2.8
3	92	3.3	4.3

* Calculated as M(PGMA) = n(GMA)/n(CPPA) × M(GMA) × conversion + M(CPPA).

The sensitivity of the surface plasmons to changes in the optical properties enables the monitoring and characterization of a metal–dielectric interface upon coating, deposition, or immobilization of organic molecules. In the Kretschmann configuration, SPR is measured from the evanescent wave propagated when a light source hits the back of the gold/glass substrate, which creates the surface plasmon wave that propagates through the metal–dielectric layer. The energy transfer between the evanescent wave to the surface plasmon wave at different angles translates to changes in reflectivity, which undergoes a sharp dip when energy transfer is complete—this is known as the resonance angle. The position of the resonance angle depends on the refractive index and thickness of the metal film [21]. Therefore, any changes on the metallic substrate can be quantified by analyzing the resulting SPR reflectivity curves and resonant angles. As can be seen from Table 3, there is significant shifting of the relative resonance angle with every modification step compared to the bare gold surface. This points to the successful immobilization of the SAM and graft polymer layer. To measure the theoretical layer thickness, reflectivity curves were simulated using Winspall 3.02 software, which is based on the Fresnel equation [11]. Using refractive index values from the literature (Table 1), resonance angle shifts can be correlated to the thickness of the SAM and grafted polymer layers (Figure 4) [22]. Shifts in reflectivity are minimal as the layer changes are on the nanometer scale (<10 nm). Results of fitting are shown in Table 3. There was close correspondence between theoretical layer thickness (taken from the literature and calculation based on bond lengths) and the values calculated from the simulation. These results will be verified by atomic force microscopy in the next phase of the study. It should also be noted that the standard deviation values of angle shifts are relatively high, which may indicate some degree of nonuniformity in the modified layer.

Table 3. Relative resonance angle from SPR measurements and calculated layer thickness using Winspall simulation and fitting.

Layer	Relative Resonance Angle (°)	Calculated Thickness (nm)	Theoretical Thickness (nm)
Gold	1.93 ± 0.33	n.a.	n.a.
DDT	2.13 ± 0.33	1.45 ± 0.18	1.7 [a]
PGMA (2 kGy)	2.63 ± 0.30	4.25 ± 0.69	4.6 [b]
PGMA (3 kGy)	3.18 ± 0.39	8.10 ± 0.43	6.9 [b]

[a] Taken from the literature [23]; [b] Calculated from bond lengths of PGMA [24] based on molecular weights from NMR analysis.

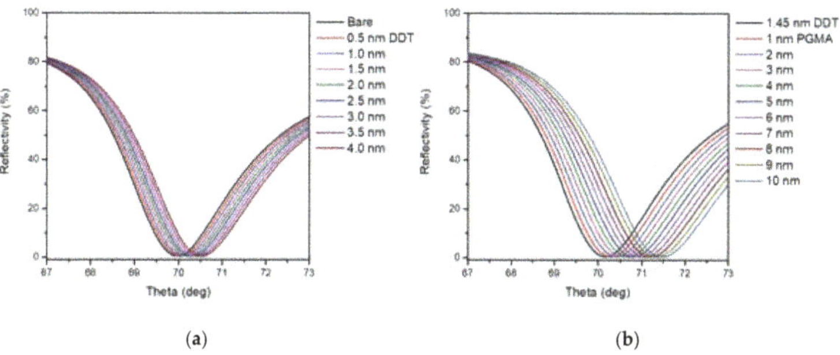

Figure 4. Calculated SPR curves for gold modified with (**a**) DDT then (**b**) PGMA.

4. Conclusions

In the endeavor to develop SPR-based sensors using a facile and environmentally friendly technique, RAFT-RIGP was employed in the modification of gold surfaces. The process involves an initial self-assembly of a thiol-based compound, whose carbon skeleton will serve as the initiation sites once exposed to ionizing radiation. This was then followed by simultaneous grafting using glycidyl methacrylate and 4-cyano-4-(phenylcarbonothioylthio) pentanoic acid in Tween 20 emulsion. Successful surface modification was verified using FTIR analysis and SPR measurement. Winspall simulation and linear fitting of layer thickness with the resonance angle indicate nanometer-scale layers were successfully immobilized. The calculated thickness values correspond well with theoretical values based on the literature. Standard deviation of resonance angles indicates some possible issues with the uniformity of the modified layers across the substrate that needs further probing. Nevertheless, the control extended by the RAFT mechanism over RIGP seems very apparent and, with some improvements, will lead to a more reproducible synthetic process. Additional characterizations and further functionalizations will be carried out in subsequent studies.

Author Contributions: Conceptualization, J.F.M., A.C.C.Y.; methodology, J.F.M., A.C.C.Y., B.J.D.B., P.N.L.H.; formal analysis, B.J.D.B., J.F.M., A.C.C.Y., P.N.L.H.; investigation, B.J.D.B., P.J.E.C., J.A.A.L., P.N.L.H.; writing—original draft preparation, B.J.D.B.; writing—review and editing, J.F.M., A.C.C.Y.; supervision, J.F.M. All authors have read and agreed to the published version of the manuscript.

Funding: This research received no external funding.

Institutional Review Board Statement: Not applicable.

Informed Consent Statement: Not applicable.

Data Availability Statement: Not applicable.

Acknowledgments: The authors acknowledge the financial support given by the International Atomic Energy Agency under contract PHI-23160 of CRP F22070 Enhancing the Beneficial Effects of Radiation Processing in Nanotechnology.

Conflicts of Interest: The authors declare no conflict of interest.

References

1. Homola, J.; Yee, S.S.; Gauglitz, G. Surface plasmon resonance sensors: Review. *Sens. Actuators B Chem.* **1999**, *54*, 3–15. [CrossRef]
2. Wijaya, E.; Lenaerts, C.; Maricot, S.; Hastanin, J.; Habraken, S.; Vilcot, J.P.; Boukherroub, R.; Szunerits, S. Surface plasmon resonance-based biosensors: From the development of different SPR structures to novel surface functionalization strategies. *Curr. Opin. Solid State Mater. Sci.* **2011**, *15*, 208–224. [CrossRef]
3. Vosloo, J.J.; De Wet-Roos, D.; Tonge, M.P.; Sanderson, R.D. Controlled free radical polymerization in water-borne dispersion using reversible addition-fragmentation chain transfer. *Macromolecules* **2002**, *35*, 4894–4902. [CrossRef]
4. Grasselli, M.; Betz, N. Electron-beam induced RAFT-graft polymerization of poly (acrylic acid) onto PVDF. *Nucl. Instrum. Methods Phys. Res. Sect. B Beam Interact. Mater. Atoms* **2005**, *236*, 201–207. [CrossRef]
5. Matyjaszewski, K.; Spanswick, J. Controlled/living radical polymerization. *Mater. Today* **2005**, *8*, 26–33. [CrossRef]
6. Barsbay, M.; Güven, O. RAFT mediated grafting of poly (acrylic acid) (PAA) from polyethylene/polypropylene (PE/PP) nonwoven fabric via preirradiation. *Polymer* **2013**, *54*, 4838–4848. [CrossRef]
7. Barsbay, M.; Güven, O. A short review of radiation-induced raft-mediated graft copolymerization: A powerful combination for modifying the surface properties of polymers in a controlled manner. *Radiat. Phys. Chem.* **2009**, *78*, 1054–1059. [CrossRef]
8. Barsbay, M.; Kodama, Y.; Güven, O. Functionalization of cellulose with epoxy groups via γ-initiated RAFT-mediated grafting of glycidyl methacrylate. *Cellulose* **2014**, *21*, 4067–4079. [CrossRef]
9. Barsbay, M.; Güven, O.; Stenzel, M.H.; Davis, T.P.; Barner-Kowollik, C.; Barner, L. Verification of controlled grafting of styrene from cellulose via radiation-induced RAFT polymerization. *Macromolecules* **2007**, *40*, 7140–7147. [CrossRef]
10. Madrid, J.F.; Ueki, Y.; Abad, L.V.; Yamanobe, T.; Seko, N. RAFT-mediated graft polymerization of glycidyl methacrylate in emulsion from polyethylene/polypropylene initiated with γ-radiation. *J. Appl. Polym. Sci.* **2017**, *134*, 45270. [CrossRef]
11. Mitsushio, M.; Masunaga, T.; Higo, M. Theoretical considerations on the responses of gold-deposited surface plasmon resonance-based glass rod sensors using a three-layer Fresnel equation. *Plasmonics* **2014**, *9*, 451–459. [CrossRef]
12. Res-Tec Surface Plasmon Resonance—The RES-TEC Quick Datasheet. Available online: http://res-tec.de/pdf/res-tec-optical-constants-database.pdf (accessed on 10 September 2021).
13. Johnson, P.B.; Christy, R.W. Optical Constants of the Noble Metals. *Phys. Rev. B* **1972**, *6*, 4370. [CrossRef]

14. Sigma-Aldrich 1-Dodecanethiol. Available online: https://www.sigmaaldrich.com/PH/en/product/aldrich/471364 (accessed on 10 September 2021).
15. Sigma-Aldrich Glycidyl Methacrylate. Available online: https://www.sigmaaldrich.com/PH/en/product/aldrich/779342 (accessed on 10 September 2021).
16. Guo, Q.; Li, F. Self-assembled alkanethiol monolayers on gold surfaces: Resolving the complex structure at the interface by STM. *Phys. Chem. Chem. Phys.* **2014**, *16*, 19074–19090. [CrossRef]
17. Sharma, A.; Singh, B.P.; Gathania, A.K. Synthesis and characterization of dodecanethiol-stabilized gold nanoparticles. *Indian J. Pure Appl. Phys.* **2014**, *52*, 93–100.
18. Chiefari, J.; Chong, Y.K.; Ercole, F.; Krstina, J.; Jeffery, J.; Le, T.P.T.; Mayadunne, R.T.A.; Meijs, G.F.; Moad, C.L.; Moad, G. Living free-radical polymerization by reversible addition−fragmentation chain transfer: The RAFT process. *Macromolecules* **1998**, *31*, 5559–5562. [CrossRef]
19. Sharif, J.; Mohamad, S.F.; Fatimah Othman, N.A.; Bakaruddin, N.A.; Osman, H.N.; Güven, O. Graft copolymerization of glycidyl methacrylate onto delignified kenaf fibers through pre-irradiation technique. *Radiat. Phys. Chem.* **2013**, *91*, 125–131. [CrossRef]
20. Perrier, S. 50th Anniversary Perspective: RAFT Polymerization—A User Guide. *Macromolecules* **2017**, *50*, 7433–7447. [CrossRef]
21. Knoll, W. Interfaces and thin films as seen by bound electromagnetic waves. *Annu. Rev. Phys. Chem.* **1998**, *49*, 569–638. [CrossRef] [PubMed]
22. Jiang, G.; Deng, S.; Baba, A.; Huang, C.; Advincula, R.C. On the Monolayer Adsorption of Thiol-Terminated Dendritic Oligothiophenes onto Gold Surfaces. *Macromol. Chem. Phys.* **2010**, *211*, 2562–2572. [CrossRef]
23. Godin, M.; Williams, P.J.; Tabard-Cossa, V.; Laroche, O.; Beaulieu, L.Y.; Lennox, R.B.; Grütter, P. Surface stress, kinetics, and structure of alkanethiol self-assembled monolayers. *Langmuir* **2004**, *20*, 7090–7096. [CrossRef]
24. Reed, K.M.; Borovicka, J.; Horozov, T.S.; Paunov, V.N.; Thompson, K.L.; Walsh, A.; Armes, S.P. Adsorption of sterically stabilized latex particles at liquid surfaces: Effects of steric stabilizer surface coverage, particle size, and chain length on particle wettability. *Langmuir* **2012**, *28*, 7291–7298. [CrossRef] [PubMed]

Proceeding Paper

The Influence of Adding a Functionalized Fluoroalkyl Silanes (PFDTES) into a Novel Silica-Based Hybrid Coating on Corrosion Protection Performance on an Aluminium 2024-t3 Alloy [†]

Magdi H. Mussa [1,2,3,*], Yaqub Rahaq [3], Sarra Takita [3], Farah D. Zahoor [3,4], Nicholas Farmilo [3,5] and Oliver Lewis [3]

[1] Mechanical and Energy Department, The Libyan Academy of Graduate Study, Tripoli P.O. Box 79031, Libya
[2] Mechanical Engineering Department, Sok Alkhamis Imsehel High Technical Institute, Sok Alkhamis, Tripoli P.O. Box 79033, Libya
[3] Material and Engineering Research Institute (MERI), Sheffield Hallam University, Sheffield S1 1WB, UK; yakoobsameer111@gmail.com (Y.R.); sarah.a.takita@gmail.com (S.T.); F.Zahoor@sheffield.ac.uk (F.D.Z.); nfarmilo@gmail.com (N.F.); O.Lewis@shu.ac.uk (O.L.)
[4] Department of Chemistry, University of Sheffield, Brook Hill, Sheffield S3 7HF, UK
[5] Tideswell Business Development Ltd., Ravensfield Sherwood Rd., Buxton SK17 8HH, UK
* Correspondence: magdimosa1976@gmail.com; Tel.: +44-7404496955
† Presented at the 2nd International Online Conference on Polymer Science—Polymers and Nanotechnology for Industry 4.0, 1–15 November 2021; Available online: https://iocps2021.sciforum.net/.

Citation: Mussa, M.H.; Rahaq, Y.; Takita, S.; Zahoor, F.D.; Farmilo, N.; Lewis, O. The Influence of Adding a Functionalized Fluoroalkyl Silanes (PFDTES) into a Novel Silica-Based Hybrid Coating on Corrosion Protection Performance on an Aluminium 2024-t3 Alloy. *Mater. Proc.* **2021**, *7*, 6. https://doi.org/10.3390/IOCPS2021-11240

Academic Editor: Shin-ichi Yusa

Published: 30 October 2021

Publisher's Note: MDPI stays neutral with regard to jurisdictional claims in published maps and institutional affiliations.

Abstract: Silica-based coatings prepared using sol-gel polymerizing technology have been shown to exhibit excellent chemical stability combined with reducing the corrosion of metal substrates, showing promising use in aerospace and marine applications to protect light alloys. Moreover, this technology is an eco-friendly technique route for producing surface coatings, showing high potential for replacing toxic pre-treatment coatings of traditional conversation chromate coatings. This study aims to investigate the enhancement in corrosion protection of a hybrid-organic-inorganic silica-based coating cured at 80 °C by increasing the hydrophobicity to work on the aluminium 2024-T3 alloy. This approach involving a novel silica-based hybrid coating was prepared by introducing 1H,1H,2H,2H-perfluorodecyltriethoxysilane (PFDTES) into the base hybrid formula created from tetraethylorthosilicatesilane (TEOS) and triethoxymethylsilane (MTMS) precursors; this formula was enhanced by introducing a Polydimethylsiloxane polymer (PDMS). The corrosion protection properties of these coatings were examined by being immersed in 3.5% NaCl with electrochemical impedance testing (EIS) and Potentiodynamic polarization scanning (PDPS). The chemical elements confirmation was performed using infrared spectroscopy (ATR-FTIR); all this was supported by analysing the surface morphology before and after the immersion by using scanning electron microscopy (SEM). The results of the electrochemical impedance testing analyses reveal the new open finite-length diffusion circuit element due to electrolyte media diffusion prevented by fluorinated groups. Additionally, it shows increases in corrosion protection arising from the increasing hydrophobicity of the fluorinated coating compared to other formulas cured under similar conditions and bare substrate. Additionally, the modified sol-gel exhibited improved resistance to cracking, while the increased hydrophobicity may also promote self-cleaning.

Keywords: silica-based hybrid sol-gel coating; electrochemical testing; corrosion protection; aluminium alloys

1. Introduction

The hybrid silica-based derived coatings have already been identified as a potential solution for the aerospace and marine industry in terms of corrosion protection [1–3]. The sol-gel process is an eco-friendly method of surface protection, which offers many options, including using single- or multi-layer coating systems with anti-corrosion and antifouling systems [2,4–6]. Additionally, sol-gel coatings can present other desirable properties, such as preventing ice accumulation, oxidation resistance and abrasion resistance [7–9]. Nevertheless, long-term exposure to moisture can negatively influence many coatings systems' adhesion and cohesion properties [4,10]. Hydrophobicity has a pivotal role in reducing the adhesion between the surfaces and direct electrolyte exposure, reducing the diffusion in coatings' pores. The precursor 1H,1H,2H,2H-Perfluorodecyltriethoxysilane (PFDTES), as shown in Figure 1, is used for hydrophobicity polyvinylidene fluoride (PVDF) to protect the metal surfaces from fouling and biocorrosion, while being reasonably cost-effective, which potentially can be used to create anti-icing and anti-corrosion surfaces [11,12].

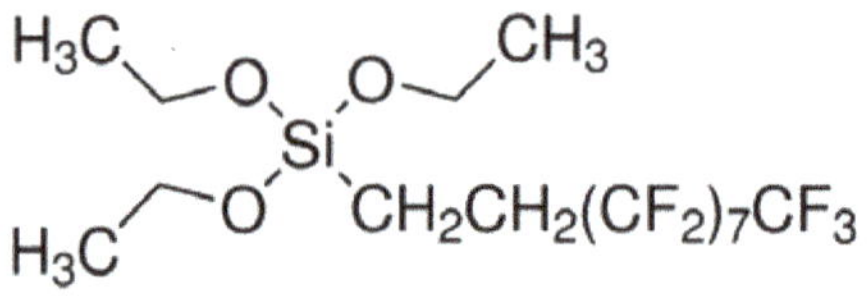

Figure 1. Chemical structure of 1H,1H,2H,2H-Perfluorodecyltriethoxysilane (PFDTES) precursor.

In this study, the performance of a functionalized silica-based sol-gel coating by introducing a 1H,1H,2H,2H-Perfluorodecyltriethoxysilane (PFDTES) precursor was investigated against corrosion by utilizing 3.5% NaCl to simulate the extended exposure similar to applications.

2. Experimental Work

2.1. Substrate Preparation

Q-panels were supplied by Q-Lab made of aluminium alloy AA2024 T3 with dimensions 102 mm × 25 mm × 1.6 mm for use as test substrates [13]. The acetone was used to remove organic residues such as grease oils or fats present on the substrate and then placed in an ultrasonic water bath for 5 min for additional cleaning with a standard alkaline solution cleaner, followed by rinsing deionized water (DI) and nitrogen drying.

2.2. Sol-Gel Preparation

The hybrid silica-based sol-gel was prepared by mixing tetraethylorthosilicate silane (TEOS), trimethoxymethyl silane (MTMS) and isopropanol (all purchased from Sigma-Aldrich (St. Louis, MO, USA)) and with dropwise additions of DI water at the molar ratio of 18:14:17:220, respectively. Then, it was enhanced by adding 12 mol% of Polydimethylsiloxane polymer (PDMS) solution. To complete the hydrolysing and the condensation polymerization reaction, the mixture was stirred with dropwise additions of nitric acid as a catalyst; the formulation was then stirred for 24 h [13]. This formula was used as the unmodified base coating and labelled as SHX. Next, the modified, fluorinated hybrid sol-gel, labelled as PF-SHX, was prepared by adding 1.5 vol.% of PFDTES from Sigma-Aldrich into the original SHX sol-gel formulation.

The formula was applied by spray coating the clean substrate and building it up over three passes. After that, the coated samples were left in the atmosphere for 10 min before being thermally annealed at 80 °C for 4 h—the chosen samples with a thickness of 16 ± 2 micrometre were chosen. Table 1 shows simple codes used to identify samples.

Table 1. Sample identification table.

No.	Identifier	Base Composite Sol-Gel	(PFDTES) *v/v*%	Curing Temperature
1.	SHX-80	TEOS + MTMS + PDMS	-	80 °C
2.	PF-SHX-80	TOES + MTMS + PDMS	1.5%	80 °C
3.	Bare AA2024 T3	-	-	-

3. Results and Discussion

3.1. Electrochemical Corrosion Testing

3.1.1. Potentiodynamic Polarization (PDPS)

Potentiodynamic polarization scans were performed on all the samples with hybrid sol-gel coatings. The results for SHX-80 and PF-SHX-80 are presented in Figure 2, along with the result of a test conducted on a bare 2024-T3 aluminium alloy for comparison. The values of corrosion potential (E_{corr}) and measured current density (I_{corr}) were obtained by extrapolating the cathodic and anodic curves using the Tafel extrapolation method. The results are shown in Table 2. The results show that the coatings reduced the measured current when compared to the uncoated aluminium alloy and shifted the corrosion potential to more noble values. This phenomenon was more apparent in the PF-SHX-80 coating, which showed a shift of 199 mV compared to the uncoated 2024-T3. The initial observation indicated that the corrosion protection offered by both sol-gel coatings is due to excellent barrier properties. Nevertheless, the shift in E_{corr} indicates that the anode is inhibited to a greater degree than the cathode; this is attributed to the fluorine-carbon atoms bridging to the substrate [14,15].

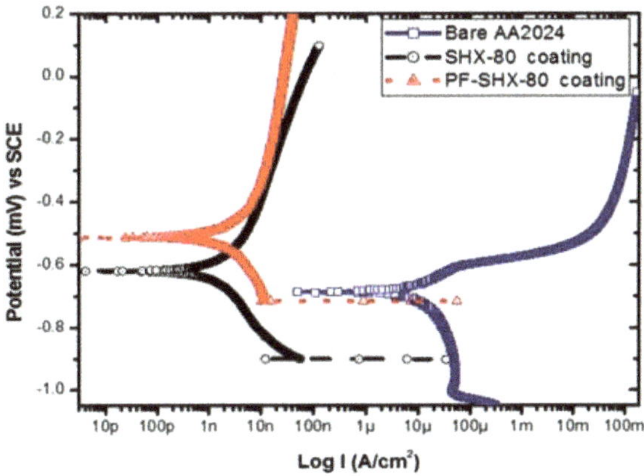

Figure 2. PDPS Polarization curves for the bare AA 2024-T3 alloy and sol-gel coated samples.

Table 2. Parameters obtained from Tafel extrapolation for bare AA 2024-T3, sol-gel coated samples.

Sample	E_{corr} [mV vs. SCE]	I_{corr} [A/cm²]	OCP [mV vs. SCE]
Bare-AA 2024	-662 ± 2	7.10×10^{-6}	-640
SHX-80 coating	-608 ± 2	1.02×10^{-9}	-708
PF-SHX-80 coating	-521 ± 2	1.22×10^{-9}	-658

3.1.2. Electrochemical Impedance Spectroscopy (EIS) Analysis

Tests were performed over a period of 14 days. Figure 3a,b show Bode impedance magnitude plots for PF-SHX-80 and the SHX-80 coated samples and bare AA2024-t3 sample in the first hour and after 336 h.

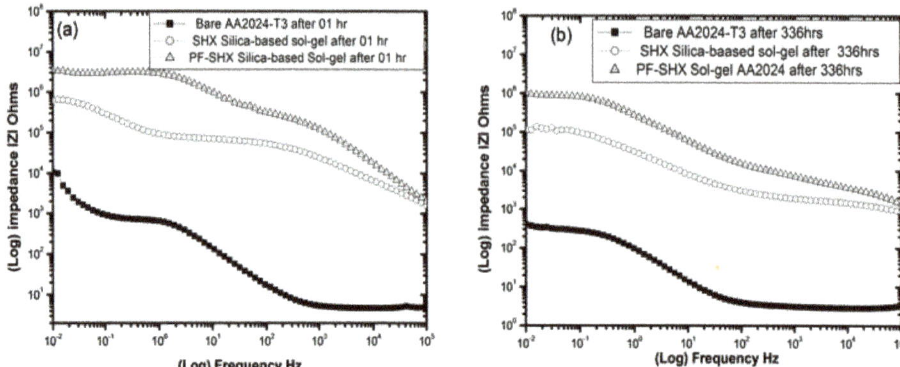

Figure 3. Bode Impedance magnitude plots for PF-SHX-80 and SHX-80 and bare samples (**a**) from 01 h and (**b**) after 336 h.

The overall impedance was increased by approximately one order of magnitude for the PF-SHX-80-coated samples compared to the SHX-80 samples, with impedance values of 6.8×10^6 and 6.8×10^5 Ω/cm^2, respectively, after one hour. At frequencies between 100 and 105 Hz, the impedance curve for the PF-SHX-80 sample reveals pure capacitive behaviour (C). Then, the impedance slowly increased from about 6.6×10^5 Ω/cm^2 in the middle range of the frequencies and finally reached a point where the rate of increase impedance is small at the low frequencies. Similarly, in the EIS measurements of the SHX-80-coated sample, a noticeable drop in impedance was observed at about 1.0×10^5 Ω/cm^2 after 14 days. On the other hand, this SHX coating still revealed higher impedance compared to the bare substrate. The increased R_{ct} values obtained from PF-SHX are consistent with the anodic inhibition obtained through a fluorine-influenced interface [15].

3.1.3. Electrochemical Equivalent Circuits Fitting for Both Sol-Gel Coatings

Tables 3 and 4 below demonstrate the fitted data obtained from the EIS spectra for the SHX-80 and the PF-SHX-80 sol-gel coatings. The equivalent circuits were used to simulate the corrosion mechanism on the coated sample in 01, 48 and 144 h. A time-constant element (Q) was used in these circuits instead of an ideal capacitor C to account for current leakage in the alternating current signals' capacitor and/or the frequency dispersion effect [14,16].

Table 3. The fitted data obtained from EIS spectra for the PF-SHX-80 sol-gel coating after various immersion times.

Sample	Element	Immersion Time (h)		
		01	48	144
	Circuit	R(Q(R(QR)))	R(Q(RO)(QR))	R(Q(RO)(RQ))
	R_s	10	18	45
	Q_{ct}	8.913×10^{-10}	9.326×10^{-10}	1.915×10^{-9}
	n	1	1	0.900
	R_{ct}	2.220×10^6	6.522×10^4	3.776×10^4
	O_{ct}	-	2.823×10^{-7}	4.866×10^{-7}
	B	-	0.469	0.618
	Q_{iL}	3.800×10^{-8}	1.103×10^{-7}	3.592×10^{-7}
	n	0.772	0.803	0.800
	R_{iL}	3.319×10^6	9.675×10^5	2.136×10^4

Table 4. The fitted data obtained from EIS spectra for the SHX-80 sol-gel coating after various immersion times.

Sample	Element	Immersion Time (h)		
		01	48	144
	Circuit	R(Q(R(QR)))	R(Q(R(Q(RW))))	R(Q(R(Q(RW))))
	R_s	100	205	195
	Q_{ct}	1.085×10^{-7}	2.059×10^{-7}	6.118×10^{-6}
	n	0.649	0.800	0.752
	R_{ct}	7.294×10^4	817	110
	Q_{iL}	4.934×10^{-6}	1.236×10^{-6}	9.815×10^{-6}
	n	0.827	0.800	0.818
	R_{iL}	7.790×10^5	3.504×10^6	1.475×10^5
	W	-	2.317×10^{-5}	8.084×10^{-5}

The elements used for the equivalent circuits were solution resistance (R_s), coating resistance (R_{ct}), coating constant phase elements (Q_{ct}), intermediate oxide layer resistance (R_{iL}), intermediate oxide layer capacitance (Q_{iL}), finite Warburg-circuit element (O) and Warburg-circuit element (W) [16].

At the first hour of immersion, both samples illustrate the same behaviour with three resistance and two time constants, respectively, as shown in Figure 4a,c. However, after 48 h of immersion of both coated samples, they started behaving individually. The SHX-80 coating's results indicate that there are three time constants. The first one arises in the high-frequency range and may be attributed to capacitive effects at the coating/aluminium/aluminium oxide interfacial layer in the coating, the second one is due to the diffusion properties of the coating, and the third one may be attributed to the Warburg-circuit element (W) as a result of diffusion to the substrate surface, as shown in Figure 4d. On the other hand, the PF-SHX-80-coated sample kept two time constants with a finite Warburg-circuit element (O) in the coating zone, as shown in Figure 4b. This (O) element is thought to originate from the hydrophobic nature of the fluorinated group ($-CH_2CH_2(CF2)_7CF_3$) from the PFDTES additive, which may prevent the diffusion of electrolyte into the fluorinated coating.

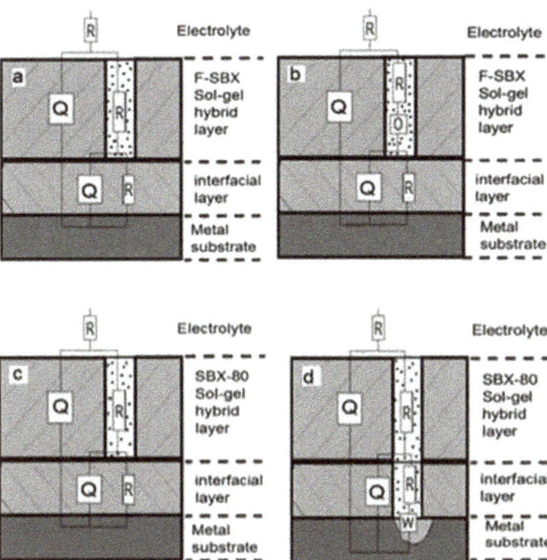

Figure 4. The modelling of (**a**,**b**) PF-SHX-80 and (**c**,**d**) SHX-80 coating systems.

3.2. Confirmation of PFDTES Addition in Sol-Gel Formula

The successful integration of the fluorinated functional groups from the (PFDTES) precursor into the sol-gel was confirmed by comparing the infrared spectrum obtained from the PF-SHX-80 coating to the unmodified SHX-80, as shown in Figure 5. The presence of C-F bonds can be confirmed in the spectral range between 1400 and 900 cm^{-1}; also, C-F$_2$ with C-F$_3$ bonds can be observed in the spectrum by the presence of bands at 1140 and 1250 cm^{-1}. The presence of the bands was highlighted with perpendicular lines on the spectrum of the based and modified sol-gels [17].

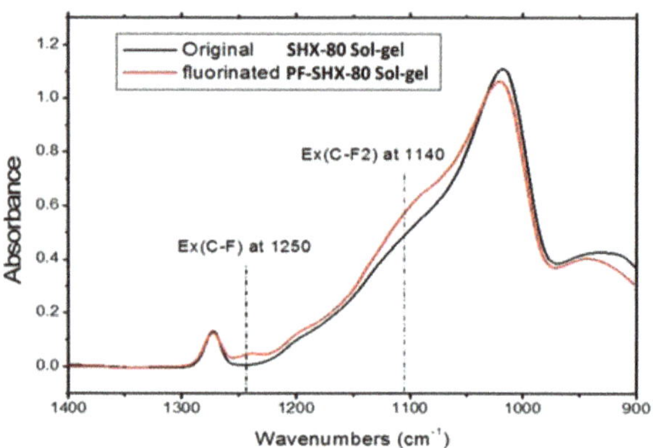

Figure 5. ATR-FTIR spectra showing the effect of PFDTES addition to the unmodified SBX-80 sol-gel.

3.3. Water Contact Angle Measurements of SHX-80- and PF-SHX-80-Coated Samples

Figure 6 shows a typical bar chart of the mean values of the water contact angle measurement droplets on both coating systems. The result of the measured average water contact angle (WCA) of the original SHX-80 coating was $67° \pm 2°$. The result of WCA on the modified PF-SHX-80 sol-gel coating was $118° \pm 3°$. The higher water contact angle recorded for the PF-SHX-80 shows that its wettability is lower than that of the SHX-80 as a result of the increased hydrophobicity of the fluorinated F-SBX-80 coating [18].

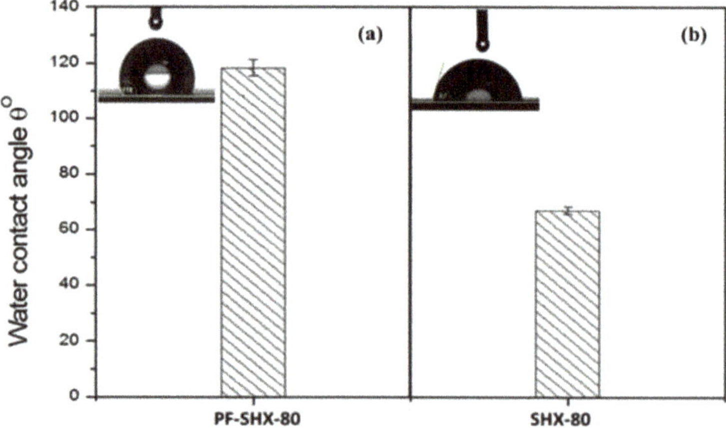

Figure 6. Bar chart showing the droplets and the mean values of WCA of (**a**) PF-SBX-80 and (**b**) SBX-80 coatings.

3.4. Scanning Electron Microscopy Imaging

Both coated samples SHX-80 and PF-SHX-80 demonstrated an ability to provide good barrier corrosion protection to the aluminium alloy substrate during long immersion. The visual examination of samples immediately after immersion showed no apparent degradation or damage to both coatings. Nevertheless, after longer immersion times (greater than ten days), the SHX-80 coating was susceptible to the formation of microcracks and pitting under the coating. The cracks were observed to be around 1–3 μm wide on the surface, as shown in the SEM images in Figure 7a. Since the presence of water under the film could lead to swelling and the loss of coating–substrate adhesion, leading to the cracking observed in the case of SHX-80. However, the PF-SHX-80 coating showed excellent resistance to cracking under similar circumstances, and this is attributable to the flexibility properties new coating, as shown in Figure 7b.

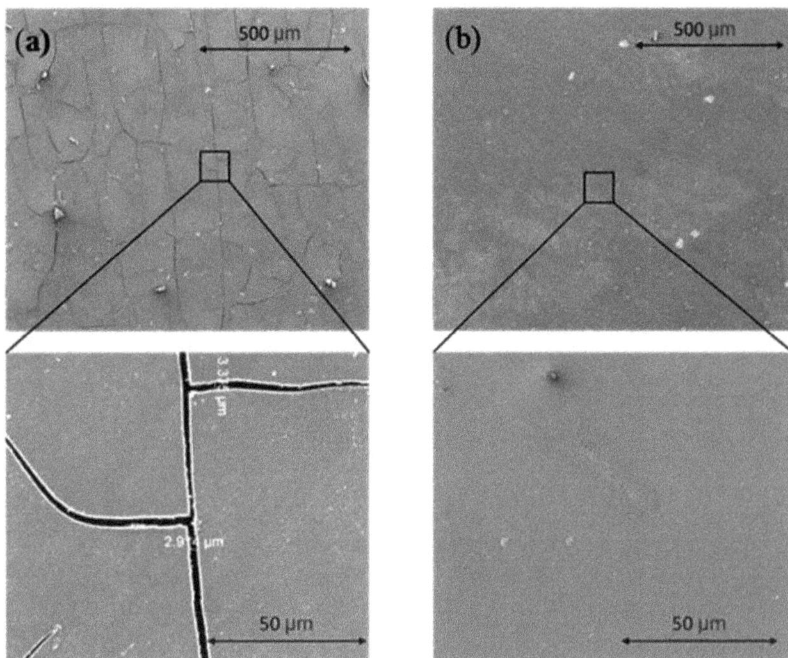

Figure 7. Secondary electron SEM micrographs of the long immersion effect on both coatings (a) SHX-80 and (b) PF-SHX-80.

4. Conclusions

The addition of the PFDTES precursor has the potential to enhance the corrosion performance of the basic TEOS and MTMS silica-based hybrid sol-gel coating on AA2024-T3 substrates. The electrochemical corrosion testing techniques confirm this by enhancing the hydrophobicity of the coating compared to other coatings with the same curing temperature of 80 °C. Furthermore, the fluorinated group from PFDTES in the hybrid organic-inorganic sol-gel coating exhibits improved post-exposure cracking resistance after prolonged immersion than the unmodified sol-gel coating. Moreover, exploiting the hydrophobic nature of the PFDTES precursor in low concentrations is potentially beneficial for applications that require self-cleaning, anti-icing or antifouling properties.

Author Contributions: Conceptualization, M.H.M.; methodology, M.H.M.; validation, M.H.M. and Y.R.; data analysis, M.H.M., S.T. and Y.R.; FTIR support, S.T. and F.D.Z.; investigation, M.H.M.; resources, M.H.M., O.L. and N.F.; writing—original draft preparation, M.H.M.; writing—review and editing, M.H.M., O.L. and N.F.; project supervision O.L. and N.F. All authors have read and agreed to the published version of the manuscript.

Funding: This research was funded by the Libyan scholarship program.

Institutional Review Board Statement: Not applicable.

Informed Consent Statement: Not applicable.

Data Availability Statement: The data are not publicly available; The data files are stored on corresponding instruments and on personal computers.

Acknowledgments: The authors would like to acknowledge the facilitating support by Sheffield Hallam University at Material and Engineering research institute MERI and also the Libyan Scholarship Program for the financial support.

Conflicts of Interest: The authors declare no conflict of interest.

References

1. Marks, D.L.; Vinegoni, C.; Bredfeldt, J.S.; Boppart, S.A. Interferometric differentiation between resonant coherent anti-Stokes Raman scattering and nonresonant four-wave-mixing processes. *Appl. Phys. Lett.* **2004**, *85*, 5787–5789. [CrossRef]
2. Detty, M.R.; Ciriminna, R.; Bright, F.V.; Pagliaro, M. Environmentally benign sol-gel antifouling and foul-releasing coatings. *Acc. Chem. Res.* **2014**, *47*, 678–687. [CrossRef] [PubMed]
3. Mussa, M.H.; Zahoor, F.D.; Lewis, O.; Farmilo, N. Developing a Benzimidazole-Silica-Based Hybrid Sol–Gel Coating with Significant Corrosion Protection on Aluminum Alloys 2024-T3. *Eng. Proc.* **2021**, *11*, 3. [CrossRef]
4. Wang, H.; Akid, R.; Gobara, M. Scratch-resistant anticorrosion sol–gel coating for the protection of AZ31 magnesium alloy via a low temperature sol-gel route. *Corros. Sci.* **2010**, *52*, 2565–2570. [CrossRef]
5. Chen, S.; Li, L.; Zhao, C.; Zheng, J. Surface hydration: Principles and applications toward low-fouling/nonfouling biomaterials. *Polymer* **2010**, *51*, 5283–5293. [CrossRef]
6. Eduok, U.; Suleiman, R.; Gittens, J.; Khaled, M.; Smith, T.J.; Akid, R.; Alia, B.; Khalile, A.; El Ali, B.; Khalil, A. Anticorrosion/antifouling properties of bacterial spore-loaded sol-gel type coating for mild steel in saline marine condition: A case of thermophilic strain of Bacillus licheniformis. *RSC Adv.* **2015**, *5*, 93818–93830. [CrossRef]
7. Lev, O.; Wu, Z.; Bharathi, S.; Glezer, V.; Modestov, A.; Gun, J.; Rabinovich, L.; Sampath, S. Sol-Gel Materials in Electrochemistry. *Chem. Mater.* **1997**, *9*, 2354–2375. [CrossRef]
8. Wang, D.; Bierwagen, G.P. Sol-gel coatings on metals for corrosion protection. *Prog. Org. Coat.* **2009**, *64*, 327–338. [CrossRef]
9. Pathak, S.S.; Khanna, S. Sol-gel nano coatings for corrosion protection. *Met. Surf. Eng.* **2012**, *12*, 304–329.
10. Zhang, S.; Xianting, Z.; Yongsheng, W.; Kui, C.; Wenjian, W. Adhesion strength of sol-gel derived fluoridated hydroxyapatite coatings. *Surf. Coat. Technol.* **2006**, *200*, 6350–6354. [CrossRef]
11. Merck 1H,1H,2H,2H-Perfluorodecyltriethoxysilane 97%. Available online: https://www.sigmaaldrich.com/GB/en/product/aldrich/658758 (accessed on 22 June 2021).
12. Cui, X.-j.; Lin, X.-z.; Liu, C.-h.; Yang, R.-s.; Zheng, X.-w.; Gong, M. Fabrication and corrosion resistance of a hydrophobic micro-arc oxidation coating on AZ31 Mg alloy. *Corros. Sci.* **2015**, *90*, 402–412. [CrossRef]
13. Mussa, M. Development of Hybrid Sol-Gel Coatings on AA2024-T3 with Environmentally Benign Corrosion Inhibitors. Ph.D. Thesis, Sheffield Hallam University, Sheffield, UK, 2020.
14. Tait, W.S. *Electrochemical Impedance Spectroscopy Fundamentals, an Introduction to Electrochemical Corrosion Testing for Practicing Engineers and Scientists*; Tait, W.S., Ed.; PairODocs Publications: Racine, WI, USA, 1994; ISBN 13-978-0966020700.
15. Kobayashi, T.; Hugel, H.M. Special issue, organo-fluorine chemical science-inventing the fluorine future. In *Applied Sciences*; Multidisciplinary Digital Publishing Institute (MDPI): Basel, Switzerland, 2012; pp. 1–283, ISBN 3906980332.
16. Yabuki, A.; Yamagami, H.; Noishiki, K. Barrier and self-healing abilities of corrosion protective polymer coatings and metal powders for aluminum alloys. *Mater. Corros.* **2007**, *58*, 497–501. [CrossRef]
17. Brassard, J.; Sarkar, D.K.; Perron, J. Fluorine Based Superhydrophobic Coatings. *Appl. Sci.* **2012**, *2*, 453–464. [CrossRef]
18. Kumar, D.; Wu, X.; Fu, Q.; Ho, J.W.C.; Kanhere, P.D.; Li, L.; Chen, Z. Development of durable self-cleaning coatings using organic-inorganic hybrid sol-gel method. *Appl. Surf. Sci.* **2015**, *344*, 205–212. [CrossRef]

Proceeding Paper

Hyaluronic Acid Hydrogel Particles Obtained Using Liposomes as Templates †

Irene Abelenda Núñez [1,*], Ramón G. Rubio [1,2], Francisco Ortega [1,2] and Eduardo Guzmán [1,2]

1   Departamento de Química Física, Facultad de Ciencias Químicas, Universidad Complutense de Madrid, Ciudad Universitaria s/n, 28040 Madrid, Spain; rgrubio@quim.ucm.es (R.G.R.); fortega@quim.ucm.es (F.O.); eduardogs@quim.ucm.es (E.G.)

2   Instituto Pluridisciplinar, Universidad Complutense de Madrid, Paseo Juan XXIII 1, 28040 Madrid, Spain

*   Correspondence: irenabel@ucm.es

†   Presented at the 2nd International Online Conference on Polymer Science—Polymers and Nanotechnology for Industry 4.0, 1–15 November 2021; Available online: https://iocps2021.sciforum.net/.

Abstract: Hydrogels (HG) are 3D networks of hydrophilic macromolecules linked by different "cross-linking points", which have as a main advantage their capacity for the adsorption of large amounts of water without any apparent dissolution. This allows hydrogels to undergo reversible swelling–shrinking processes upon the modification of the environmental conditions (pH, ionic strength or temperature). This stimuli-responsiveness and their ability for entrapping in their interior different types of molecules makes hydrogels suitable platforms for drug delivery applications. Furthermore, HGs exhibit certain similarities to the extracellular tissue matrix and can be used as a support for cell proliferation and migration.

Keywords: hydrogels; hyaluronic acid; swelling; shrinkage; liposomes; template-assisted

Citation: Núñez, I.A.; Rubio, R.G.; Ortega, F.; Guzmán, E. Hyaluronic Acid Hydrogel Particles Obtained Using Liposomes as Templates. *Mater. Proc.* **2021**, *7*, 7. https://doi.org/10.3390/IOCPS2021-11222

Academic Editor: Shin-ichi Yusa

Published: 25 October 2021

Publisher's Note: MDPI stays neutral with regard to jurisdictional claims in published maps and institutional affiliations.

1. Introduction

Template-assisted methodologies have been successfully applied on the preparation of particles of alginate, agarose, milk protein or whey protein with a well-defined shape and size [1–4]. The particles obtained by templating techniques can be exploited for the encapsulation of different actives with interest in different industries and technological fields [3–5]. Among the templates used, the droplets of water in oil (W/O) emulsions are probably counted as the most extended. However, the use of emulsions present an important drawback related with the presence of an organic solvent, which can remain partially trapped in the particle matrix. This may alter the safety of the obtained particles, especially when they are intended for applications involving the interaction between particles and biological systems.

The use of liposomes as template instead of emulsion droplets for liposomes emerges as a very important alternative for reducing the use of organic compounds, and improving the toxicological profile of the obtained particles. Liposomes are defined as spherical structures consisting on a lipid bilayer surrounding a hydrophilic core, commonly filled by water, which can be a very suitable environment for preparing hydrophilic polymeric particles [6,7].

This work is focused on the preparation of hydrogels particles of hyaluronic acid. Polymer hydrogels, or simply hydrogels (HG), are three-dimensional macromolecular network, formed by hydrophilic polymer chains linked via physical or chemical interactions through different "crossing points", i.e., they form cross-linked structures. The branched structures of the HGs allow them to absorb large amounts of water, while the cross-linking of the networks prevents their dissolution (see Figure 1 for a sketch). In contact with water, these materials have the ability to swell and form elastic, soft and flexible materials, while also retaining a significant amount of solvent within their structure. These properties provide HGs with certain similarities to the extracellular tissue matrix, allowing their use as substrates for cell proliferation and migration or to control drug release [8].

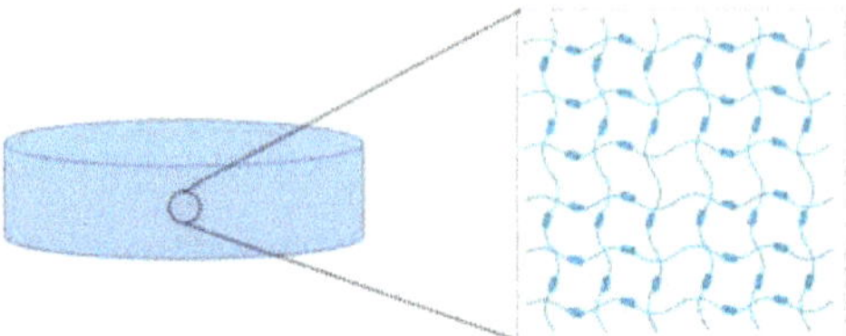

Figure 1. Sketch of the typical cross-linked structure of a polymeric hydrogel.

Hyaluronic acid (HA) is a polysaccharide frequently used in different biotechnological applications due to its natural, biodegradable, and nontoxic character which enables the formation of inert nanoparticles. In recent years, an extensive research on polysaccharide nanoparticles for several applications has been developed [9,10]. HA hydrogels have currently a large number of applications in biomaterials, including their use in tissue regeneration because of their high biocompatibility, or as drug delivery systems because their ability to retain liquids and bioactive compounds [11].

The aim of this work the fabrication of agarose nanoparticles using liposomes as templates for substituting the commonly use emulsions. The use of liposomes as templates is preferred due to their stability, simplicity of preparation and the absence of organic solvents. Furthermore, nanosized liposomes provide a suitable environment for the fabrication of nanosized hydrophilic particles.

2. Experimental Section

2.1. Chemicals

L-α-Phosphatidylcholine (PC, 2-linoleoyl-1-palmitoyl-sn-glycero-3-phosphocholine) with a purity higher than the 95% and a molecular weight of 782.08 g/mol was supplied for Alfa Aesar (Haverhill, MA, USA). Sodium hyaluronate (HANa) with molecular weight in the range 1.5–1.8×10^3 kDa was purchased from Sigma-Aldrich (Saint-Louis, MO, USA). PC and HANa were used as received without any further purification.

Hydrochloride acid (HCl, aqueous solution at 35 wt%) for fixing the pH, glucose (purity 99%) and diethyl ether (CHROMASOLV™, for High Performance Liquid Chromatography, purity 99.9%) were supplied for Sigma-Aldrich (Saint Louis, MO, USA).

Ultrapure deionized water used for cleaning and solution preparation was obtained by a multi-cartridge purification system aquaMAX™-Ultra 370 Series (Young Lin Instrument, Co., Ltd., Anyang, Korea). The water used had a resistivity higher than 18 MΩ·cm, and a total organic content lower than 6 ppm.

2.2. Preparation of Liposomes Loaded with Hyaluronic Acid

Liposomes were prepared following a procedure adapted from that commonly followed in the reverse phase evaporation method [12,13]. This technique relies on the formation of reverse micelles, which are later transformed in liposomes. For this purpose, an organic phase composed of phosphatidylcholine dissolved diethyl ether and an aqueous phase corresponding containing the substance to be encapsulated, i.e., hyaluronic acid, at a concentration of 0.2 g/L are mixed in 1:1 volume ratio. These mixtures are left for equilibration during 30 min, and then is centrifuged for 30 min at 400 rpm, which evidences a clear separation of phases between the aqueous and organic ones. This allows for removal of the aqueous fraction containing the excess of non-encapsulated hyaluronic acid. Afterwards, the organic fraction containing reverse micelles loaded with hyaluronic acid is mixed with water in 1:1 volume ratio, which is followed by the addition of volume similar to that added of water of an aqueous solution containing 5 wt% glucose solution. The above mixture is placed in an ultrasonic bath for 5 min, and then the organic solvent is removed using a rotary evaporator, which results in an aqueous dispersion of liposomes loaded with hyaluronic acid. It should be noted that the preparation of bare liposomes, without hyaluronic acid, was performed following a similar approach without adding hyaluronic acid to the aqueous phase.

2.3. Preparation of Hyaluronic Acid Hydrogels

Hyaluronic acid has the ability to undergo self-crosslinking process in acid medium, i.e., it can form ester bonds with another hyaluronic acid molecules. This requires to reduce the pH of the dispersion of liposomes loaded with hyaluronic acid by adding HCl down to a value of 1.5 [14]. Thus, it is possible to form inter- and intra-chain ester bonds, which leads to the formation of hydrogels particles adopting the form of the environment containing the polymer chains. Figure 2 shows a sketch representing the formation on an ester bonds between two hyaluronic acid monomers.

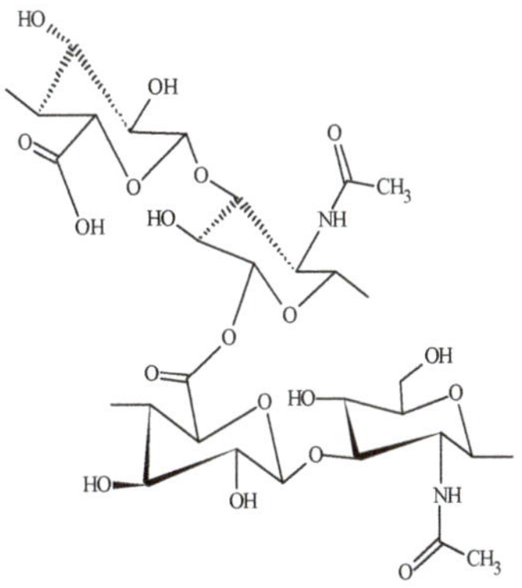

Figure 2. Sketch of the cross-linking between hyaluronic acid molecules.

2.4. Characterization Techniques

Dynamic Light Scattering (DLS) experiments for characterizing the size of bare liposomes, hyaluronic acid loaded liposomes and hyaluronic acid hydrogels were performed by using a Zetasizer Nano ZS device (Malvern Instrument, Ltd., Malvern, UK). Thus, it is possible to perform the evaluation of the size of the object dispersed in the aqueous medium in terms of the apparent hydrodynamic diameter (in the following diameter, d).

The cross-linking of hyaluronic acid was confirmed by measuring the changes in the infrared spectrum by using Spectrophotometer FT-IR Nicolet iS50 (Thermo Fisher Scientific, Waltham, MA, USA).

3. Results and Discussion

3.1. Verification of the Formation of Liposomes Loaded with Hyaluronic Acid Using the Reverse Phase Technique

The use of DLS has provided information about the formation of liposomes by using the reverse phase technique, and the monodisperse character of the obtained liposomes. Figure 3 shows the size distribution obtained for the liposomes contained in a dispersion obtained following the above describe procedure. From the results, it is clear that the used methodology allows for obtaining liposomes with sizes contained in a very narrow diameter distribution (monodisperse dispersions), and an average diameter of 206 ± 2 nm.

The possibility to obtain monodisperse liposomes is key for their use as templates for obtaining hydrogels with controlled sizes and shapes.

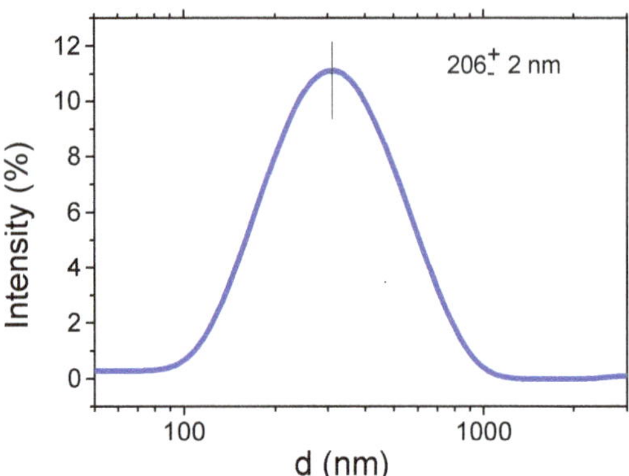

Figure 3. Distribution of diameters obtained by DLS for liposomes loaded with hyaluronic acid.

3.2. Hydrogel Formation

The ability of hyaluronic acid to undergo a self-crosslinking process under acidic conditions was stated above. This allows tuning the degree of crosslinking of the hydrogels by varying the HCl concentration added to the aqueous medium. The effect of the degree of crosslinking as function of the added HCl concentration is reflected in changes on the average diameter of the liposomes loaded with hyaluronic acid as is shown in Figure 4. The increase in HCl concentration causes a swelling of the liposomes, which can be rationalized in terms of changes on the crosslinking of the HA encapsulate. This leads to the formation of hydrogel particles with different rigidity and swelling degree, which push the liposomes walls, resulting in an increase in the average thickness of the liposomes. Therefore, by modifying the pH of the medium, it is possible to obtain liposomes with hyaluronic acid hydrogels that have different degrees of crosslinking.

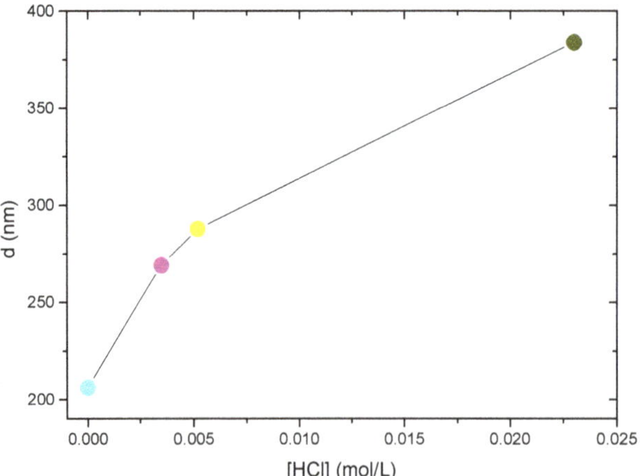

Figure 4. Average diameter for liposomes loaded with hyaluronic acid as function of the HCl in the aqueous medium.

It should be noted that an excessive increase in the HCl concentration in the medium results in the degradation of the liposomes, which results in a release of the encapsulated hydrogel particles followed by their aggregation, as evidenced in the images shown in Figure 5.

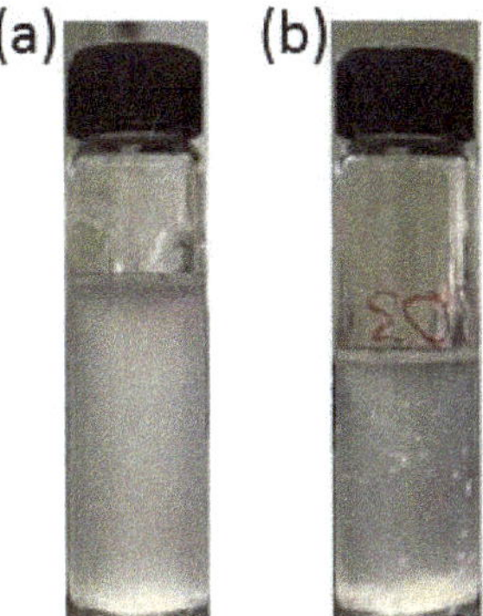

Figure 5. Images showing a dispersion of liposomes loaded with hyaluronic acid at physiological pH (**a**) and a dispersion of liposomes and aggregate hydrogel particles at pH 2.03 (**b**).

3.3. Evaluation of the Cross-Linking of the Obtained Hydrogels

The crosslinking of hyaluronic acid hydrogel particles upon exposure at acid medium was confirmed by using infrared spectroscopy. For this purpose, the IR spectra of hyaluronic acid particles encapsulated in the liposomes and bare hyaluronic acid were analyzed by infrared spectroscopy, and the results were compared (see Figure 6). As can be observed, the sample of crosslinked hyaluronic acid presents an adsorption band at 1730 cm^{-1}, associated with the C-O tension of ester groups, which confirms the esterification process and consequently provides evidences of the formation of the hyaluronic acid hydrogels by a self-crosslinking process in acid medium.

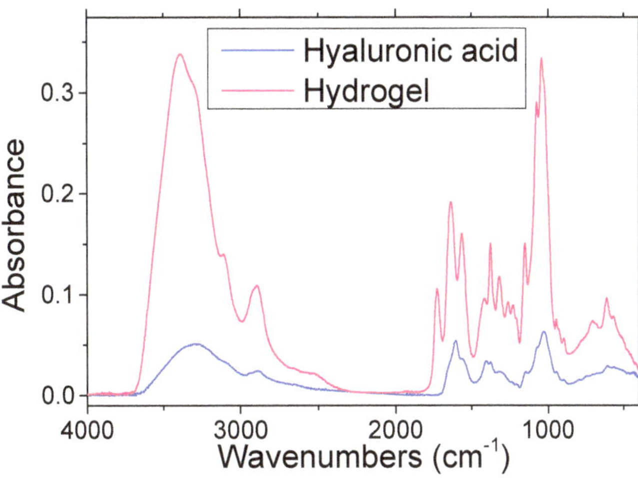

Figure 6. Infrared spectra for bare hyaluronic acid (blue line) and hyaluronic acid hydrogel (red line).

4. Conclusions

This work reports a new route for preparation of stable and monodisperse nanosized hyaluronic hydrogel nanoparticles. This was possible due to a pH-triggered gelation of the aqueous interior of the liposomes used as a template for controlling the size and morphology of the nanoparticles. Such nanoparticles can be transferred to aqueous medium to obtain a dispersion of nanoparticles which present a physico-chemical behavior reminiscent of a chemically cross-linked gel, which can undergo a reversible swelling–shrinking process. This open new avenues for design system for controlled loading and release of active molecules.

Supplementary Materials: The following supporting information can be downloaded at: https://www.mdpi.com/article/10.3390/IOCPS2021-11222/s1.

Author Contributions: Conceptualization, R.G.R. and E.G.; methodology, I.A.N.; software, I.A.N.; validation, F.O., R.G.R. and E.G.; formal analysis, I.A.N.; investigation, I.A.N., F.O., R.G.R. and E.G.; resources, F.O. and R.G.R.; data curation, I.A.N. and E.G.; writing—original draft preparation, I.A.N. and E.G.; writing—review and editing, I.A.N., F.O., R.G.R. and E.G.; visualization, I.A.N. and E.G.; supervision, R.G.R. and E.G.; project administration, E.G.; funding acquisition, F.O., R.G.R. and E.G. All authors have read and agreed to the published version of the manuscript.

Funding: This work was funded in part by MICINN (Spain) under grant PID2019-106557GB-C21, by Banco Santander-Universidad Complutense grant PR87/19-22513 (Spain) and by E.U. on the framework of the European Innovative Training Network-Marie Sklodowska-Curie Action NanoPaint (grant agreement 955612).

Institutional Review Board Statement: Not applicable.

Informed Consent Statement: Not applicable.

Data Availability Statement: Data are available upon request.

Acknowledgments: Authors acknowledge the Centro de Espectroscopía y Correlación (UCM) for the use of their facilities.

Conflicts of Interest: The authors declare no conflict of interest. The funders had no role in the design of the study; in the collection, analyses, or interpretation of data; in the writing of the manuscript, or in the decision to publish the results.

References

1. Reis, C.P.; Neufeld, R.J.; Vilela, S.; Ribeiro, A.J.; Veiga, F. Review and current status of emulsion/dispersion technology using an internal gelation process for the design of alginate particles. *J. Microencaps.* **2006**, *23*, 245–257. [CrossRef] [PubMed]
2. Matalanis, A.; Jones, O.G.; McClements, D.J. Structured biopolymer-based delivery systems for encapsulation, protection, and release of lipophilic compounds. *Food Hydrocolloids* **2011**, *25*, 1865–1880. [CrossRef]
3. Chen, L.; Subirade, M. Alginate–whey protein granular microspheres as oral delivery vehicles for bioactive compounds. *Biomaterials* **2006**, *27*, 4646–4654. [CrossRef] [PubMed]
4. Argudo, P.G.; Guzmán, E.; Lucia, A.; Rubio, R.G.; Ortega, F. Preparation and Application in Drug Storage and Delivery of Agarose Nanoparticles. *Int. J. Polym. Sci.* **2018**, *2018*, 7823587. [CrossRef]
5. Heidebach, T.; Först, P.; Kulozik, U. Microencapsulation of probiotic cells by means of rennet-gelation of milk proteins. *Food Hydrocolloids* **2009**, *23*, 1670–1677. [CrossRef]
6. Akbarzadeh, A.; Rezaei-Sadabady, R.; Davaran, S.; Joo, S.; Zarghami, N.; Hanifehpour, Y.; Samiei, M.; Kouhi, M.; Nejati-Koshki, K. Liposome: Classification, preparation, and applications. *Nanoscale Res. Lett.* **2013**, *8*, 102. [CrossRef] [PubMed]
7. Malam, Y.; Loizidou, M.; Seifalian, A.M. Liposomes and nanoparticles: Nanosized vehicles for drug delivery in cancer. *Trends Pharm. Sci.* **2009**, *30*, 592–599. [CrossRef] [PubMed]
8. Yin, S.; Cao, Y. Hydrogels for Large-Scale Expansion of Stem Cells. *Acta Biomaterialia* **2021**, *128*, 1–20. [CrossRef] [PubMed]
9. Salatin, S.; Barar, J.; Barzegar-Jalali, M.; Adibkia, K.; Milani, M.A.; Jelvehgari, M. Hydrogel nanoparticles and nanocomposites for nasal drug/vaccine delivery. *Arch. Pharm. Res.* **2016**, *39*, 1181–1192. [CrossRef] [PubMed]
10. Saboktakin, M.R.; Tabatabaee, R.M.; Maharramov, A.; Ramazanov, M.A. Design and characterization of chitosan nanoparticles as delivery systems for paclitaxel. *Carbohydr. Polym.* **2010**, *82*, 466–471. [CrossRef]
11. Kopeček, J.; Yang, J. Hydrogels as smart biomaterials. *Polym. Int.* **2007**, *56*, 1078–1098. [CrossRef]
12. Cortesi, R. Preparation of liposomes by reverse-phase evaporation using alternative organic solvents. *J. Microencaps.* **1999**, *16*, 251–256. [CrossRef] [PubMed]
13. Szoka, F.; Papahadjopoulos, D. Procedure for preparation of liposomes with large internal aqueous space and high capture by reverse-phase evaporation. *Proc. Natl. Acad. Sci. USA* **1978**, *75*, 4194–4198. [CrossRef] [PubMed]
14. Collins, M.N.; Birkinshaw, C. Physical properties of crosslinked hyaluronic acid hydrogels. *J. Mat. Sci. Mat. Med.* **2008**, *19*, 3335–3343. [CrossRef] [PubMed]

materials proceedings

Abstract

Antitumor Cytokine DR5-B-Conjugated Polymeric Poly(*N*-vinylpyrrolidone) Nanoparticles with Enhanced Cytotoxicity in Human Colon Carcinoma 3D Cell Spheroids [†]

Anne Yagolovich [1,*], Andrey Kuskov [2], Pavel Kulikov [3], Leily Kurbanova [1], Anastasia Gileva [1] and Elena Markvicheva [1]

[1] Shemyakin-Ovchinnikov Institute of Bioorganic Chemistry of the Russian Academy of Sciences, 117997 Moscow, Russia; leyli.kurbanova_1997@mail.ru (L.K.); sumina.anastasia@mail.ru (A.G.); lemarkv@hotmail.com (E.M.)

[2] Department of Biomaterials, Mendeleev University of Chemical Technology of Russia, 125047 Moscow, Russia; ankuskov@gmail.com

[3] Center of Strategic Planning and Management of Medical and Biological Health Risks, 119121 Moscow, Russia; p.kulikov.p@gmail.com

[*] Correspondence: anneyagolovich@gmail.com

[†] Presented at the 2nd International Online Conference on Polymer Science—Polymers and Nanotechnology for Industry 4.0, 1–15 November 2021; Available online: https://iocps2021.sciforum.net/.

Citation: Yagolovich, A.; Kuskov, A.; Kulikov, P.; Kurbanova, L.; Gileva, A.; Markvicheva, E. Antitumor Cytokine DR5-B-Conjugated Polymeric Poly(*N*-vinylpyrrolidone) Nanoparticles with Enhanced Cytotoxicity in Human Colon Carcinoma 3D Cell Spheroids. *Mater. Proc.* **2021**, *7*, 8. https://doi.org/10.3390/IOCPS2021-11281

Academic Editor: Marina Arrieta

Published: 1 November 2021

Publisher's Note: MDPI stays neutral with regard to jurisdictional claims in published maps and institutional affiliations.

Abstract: Self-assembled nanoparticles based on amphiphilic poly(*N*-vinylpyrrolidone) (Amph-PVP) were proposed earlier as a new drug delivery system. In the current work, we study the antitumor activity of Amph-PVP-based self-assembled polymeric micelles covalently conjugated with the antitumor receptor-specific TRAIL variant DR5-B (P-DR5-B). The Amph-PVP polymer was synthesized by the earlier developed one-step technique (Kulikov et al., Polym. Sci. Ser. D, 2017). To stabilize Amph-PVP associates, the hydrophobic core was loaded with the model substance prothionamide. For the covalent conjugation with DR5-B, the hydrophilic ends of polymeric chains were modified with maleimide, and a DR5-B N-terminal amino acid residue valine was mutated to cysteine (DR5-B/V114C). DR5-B/V114C was conjugated to the surface of polymeric micelles by the selective covalent interaction of N-terminal cysteine residue with maleimide on Amph-PVP. The cytotoxicity of DR5-B-conjugated Amph-PVP polymeric nanoparticles was investigated in 3D multicellular tumor spheroids (MCTS) of human colon carcinoma HCT116 and HT29 cells, generated by the RGD-induced self-assembly technique (Akasov et al., Int. J. Pharm., 2016). In DR5-B-sensitive HCT116 MCTS, the P-DR5-B activity slightly increased compared to that of DR5-B. However, in DR5-B-resistant HT29 MCTS, P-DR5-B significantly surpassed DR5-B in the antitumor activity. Thus, the conjugation of DR5-B with the Amph-PVP nanoparticles enhanced its tumor-cell killing capacity. In the current study, we obtain a new nano-scaled delivery system based on Amph-PVP self-aggregates coated with covalently conjugated antitumor DR5-specific cytokine DR5-B. P-DR5-B overcomes DR5-B-resistance of the human colon carcinoma MCTS in vitro. This makes Amph-PVP polymeric nanoparticles a prospective and versatile nano-scaled delivery system for the targeted proteins.

Keywords: amphiphilic polymeric nanoparticles; poly(*N*-vinylpyrrolidone); antitumor therapy; colon carcinoma; receptor-specific TRAIL variant DR5-B

Supplementary Materials: The poster presentation is available online at https://www.mdpi.com/article/10.3390/IOCPS2021-11281/s1.

Author Contributions: Conceptualization, A.Y. and A.K.; methodology, A.Y., A.K. and E.M.; validation, A.Y. and A.G.; formal analysis, A.Y., A.K., A.G. and E.M.; investigation, P.K. and L.K.; resources, A.K. and E.M.; data curation, A.Y. and A.G.; writing—original draft preparation, A.Y. and P.K.;

writing—review and editing, A.K. and E.M.; visualization, A.Y., P.K. and A.G.; supervision, E.M.; project administration, A.Y. and E.M.; funding acquisition, A.Y. All authors have read and agreed to the published version of the manuscript.

Funding: This research was funded by Russian Science Foundation grant No. 21-14-00224, https://rscf.ru/project/21-14-00224/ (protein expression and purification and development of 3D tumor models were held at Shemyakin-Ovchinnikov Institute of Bioorganic Chemistry of the Russian Academy of Sciences) and by the Russian Foundation for Basic Research grant № 18-34-00812 (nanoparticle synthesis and characterization were held at Department of Biomaterials, D. Mendeleev University of Chemical Technology of Russia and Federal State Budgetary Institution "Centre for Strategic Planning and Management of Biomedical Health Risks" of the Federal Medical Biological Agency, Moscow, Russia).

Institutional Review Board Statement: Not applicable.

Informed Consent Statement: Not applicable.

Data Availability Statement: The data presented in this study are available on request from the corresponding author.

Conflicts of Interest: The authors declare no conflict of interest.

materials proceedings

MDPI

Proceeding Paper

Light-Driven Integration of Graphitic Carbon Nitride into Polymer Materials [†]

Cansu Esen * and Baris Kumru *

Department of Colloid Chemistry, Max Planck Institute of Colloids and Interfaces, Am Mühlenberg 1, 14476 Potsdam, Germany

* Correspondence: Cansu.esen@mpikg.mpg.de (C.E.); Baris.kumru@mpikg.mpg.de (B.K.); Tel.: +49-(331)-567-9536 (C.E.); +49-(331)-567-9569 (B.K.)

† Presented at the 2nd International Online Conference on Polymer Science—Polymers and Nanotechnology for Industry 4.0, 1–15 November 2021; Available online: https://iocps2021.sciforum.net/.

Abstract: As a metal-free polymeric semiconductor with an absorption in the visible range, carbon nitride has numerous advantages for photo-based applications spanning hydrogen evolution, CO_2 reduction, ion transport, organic synthesis and organic dye degradation. The combination of g-C_3N_4 and polymer networks grants mutual benefit for both platforms, as networks are upgraded with photoactivity or formed by photoinitiation, and g-C_3N_4 is integrated into novel applications. In the present contribution, some of the recently published projects regarding g-C_3N_4 and polymeric materials will be highlighted. In the first study, organodispersible g-C_3N_4 were incorporated into a highly commercialized porous resin called poly(styrene-co-divinylbenzene) through suspension photopolymerization, and performances of resulting beads were investigated as recyclable photocatalysts. In the other study, g-C_3N_4 nanosheets were embedded in porous hydrogel networks, and so-formed hydrogels with photoactivity were transformed either into a 'hydrophobic hydrogel' or pore-patched materials via secondary network introduction, where both processes were accomplished via visible light. Since g-C_3N_4 is an organic semiconductor exhibiting sufficient charge separation under visible light illumination, a novel method for the oxidative photopolymerization of EDOT was successfully accomplished. As a result of the absence of dissolved anions during polymerization, so-formed neutral PEDOT is a highly viscous liquid that can be processed and post-doped easily, and grants facile coating processes.

Keywords: photopolymerization; photochemistry; porous materials; soft materials; hydrogels; conducting materials; organic semiconductors

Citation: Esen, C.; Kumru, B. Light-Driven Integration of Graphitic Carbon Nitride into Polymer Materials. *Mater. Proc.* **2021**, *7*, 9. https://doi.org/10.3390/IOCPS2021-11590

Academic Editor: Shin-ichi Yusa

Published: 5 November 2021

Publisher's Note: MDPI stays neutral with regard to jurisdictional claims in published maps and institutional affiliations.

1. Introduction

In this proceeding, recently published projects based on the combination of metal-free semiconductor graphitic carbon nitride (g-C_3N_4) and polymers will be summarized. The proceeding will briefly focus on how graphitic carbon nitride is successfully integrated into different reaction environments, such as in dispersion, or as a heterogeneous photo initiator, to evolve the properties and synthesis of polymer materials.

A relatively old material, g-C_3N_4 is a polymeric semiconductor discovered in 2009. Its synthesis is based on thermal polymerization (around 550 °C) of abundant and nitrogen-rich molecules, such as urea and melamine, that result in a semiconductor with photoactivity in the visible and UV range [1]. Its thermal reaction mechanism can be considered as recondensation of decomposed gaseous molecules of precursors to form a thermodynamically stable matter at elevated temperatures. It is different than a typical solid state condensation reaction; therefore, reaction conditions such as temperature ramp and applied atmosphere play a curial role in forming the desired and reproducible g-C_3N_4. Since it is a type of polymerization, the condensation degree can be also tuned by the variability of reaction conditions (mainly temperature). Moreover, further modifications, such as edge functioning and tuning repeating motifs, are just some possibilities to form diverse g-C_3N_4's. g-C_3N_4 exhibits high stability in acidic and basic environments, as well as thermal stability up to 620 °C.

Selected precursor and preparation methods directly affect all final properties, such as surface area, morphology, surface charge, crystallinity and optical properties. Photophysical features of the resulting g-C_3N_4 may vary depending on the performed techniques. Synthesized g-C_3N_4s in powder form demonstrates a slightly negative surface charge; however, the synthesis of g-C_3N_4 thin films and membranes are known in the literature as well [2,3]. Furthermore, as a typical semiconductor, g-C_3N_4 exhibits reductive and oxidative pathways upon photo-formed electrons and holes, which are a vital role of photoredox-based science. g-C_3N_4 has been a haven for photocatalysis, and has been integrated into many appealing applications by acting as a heterogeneous photocatalyst [4]. Furthermore, under light irradiation, it can form radical species which can be harnessed to conduct radical photopolymerization techniques. From a polymer perspective, g-C_3N_4 was employed as a heterogeneous photoinitator for free radical and controlled radical polymerization methodologies.

However, some drawbacks based on dominant π-π interactions between g-C_3N_4 sheets cause problems, such as processability and dispersion preparation, which hinder its further possibilities. Recently, the surface modification of g-C_3N_4 materials introduced dispersible g-C_3N_4 colloids which, addressed both in aqueous and organic environments, are attractive for new possibilities [5,6]. There are numerous well-established techniques that deal with the integration of g-C_3N_4 colloids in soft materials, membranes and polymer particles to form hybrid materials. Yet, there are still many materials to be explored that are based on g-C_3N_4 and polymer materials.

Herein, the following proceeding will highlight three recently published reports based on g-C_3N_4 and polymer combination. It will start with an organic dispersion of g-C_3N_4 to form crosslinked porous polymer beads named poly(styrene-co-divinylbenzene) through suspension polymerization, to sculpt photoactive bead materials. Aqueous dispersion of g-C_3N_4 will be employed for porous hydrogel synthesis, and the embedded g-C_3N_4 nanosheets will act as anchoring points for further light-induced modifications, i.e., pore substructuring and the synthesis of 'hydrophobic hydrogel'. Finally, g-C_3N_4 will be utilized as a heterogeneous photoinitiator to polymerize 3,4-ethylenedioxythiophene (EDOT) to form a non-doped oligoEDOT material that is prone to processing prior to doping.

2. Results

2.1. Upgrading Poly(styrene-co-divinylbenzene) Beads through Suspension Photopolymerization via Carbon Nitride

Metal-free semiconductor graphitic carbon nitride (g-C_3N_4) can act as a heterogeneous photoredox polymer initiator. This study introduces g-C_3N_4 integration into the inner surface (interfaces) of porous poly(styrene-co-divinylbenzene) beads via one-pot suspension photopolymerization (Scheme 1).

Regarding suspension polymerization principles, reaction conditions such as crosslinking ratio (25, 35, 50 wt.% divinylbenzene), the presence of porogens (toluene as co-solvent, 1-octanol as dispersion agent of vTA-CMp), and mechanical agitation (slow and medium) are investigated. According to results, 25 wt.% divinylbenzene, the presence of toluene along with 1-octanol, and medium agitation speed, were elucidated as optimal conditions.

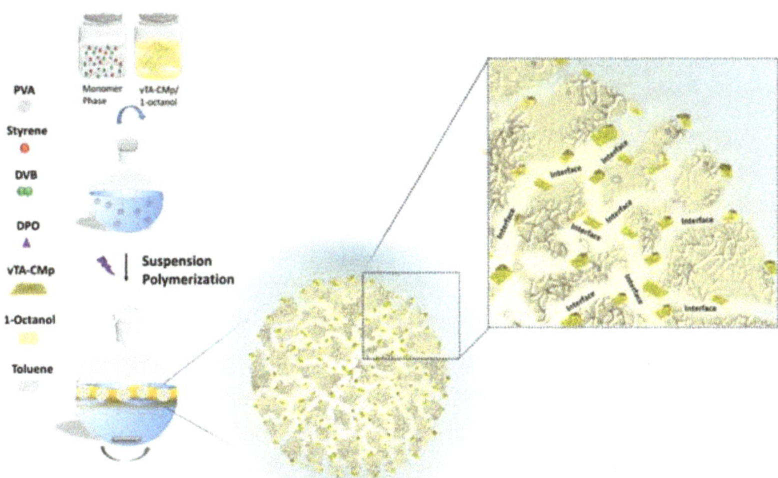

Scheme 1. Schematic overview of organomodified g-C$_3$N$_4$ (vTA-CMp) incorporation into PS-DVB beads via suspension photopolymerization [7]. Copyright 2021, John Wiley& Sons.

As shown in Figure 1, bead morphology was investigated via SEM, and the resulting images revealed a highly porous structure (Figure 1(a1,a2)). In addition, the EDX investigation exhibiting homogeneous nitrogen distribution indicated successful vTA-CMp corporation, since none of the reaction ingredients, except vTA-CMp, bear nitrogen (Figure 1b).

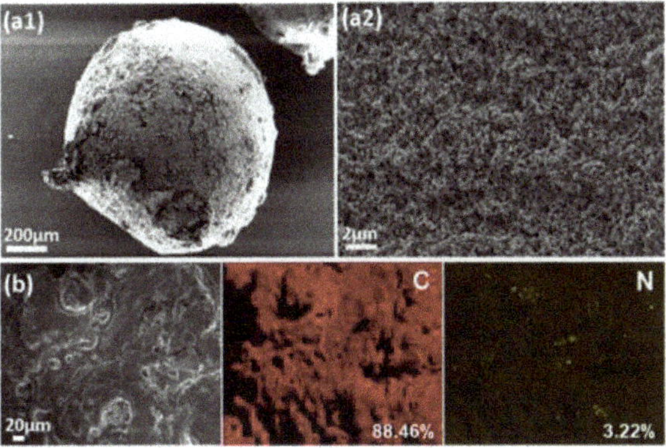

Figure 1. Scanning electron microscopy (SEM) images (**a1,a2**) of a model bead and Elemental mapping of model bead via EDX (**b**) [7]. Copyright 2021, John Wiley& Sons.

Photocatalytic properties of the selected beads were investigated by performing an aqueous Rhodamine B dye degradation experiment under visible light illumination and dark conditions (Figure 2a,b). Among the as-synthesized beads, the g-C$_3$N$_4$ photo-initiated bead (CBT5m) exhibited 89% degradation efficiency in 5 h, within the R^2 value of 0.971 (Figure 2c). Moreover, cyclic photocatalytic cycles ran up to 7 cycles and resulted in moderate loss of activity, 55.6% (Figure 2d). Lastly, the pH effect on photodegradation explicated the stability of the applied beads with no host degradation. In addition, dye adsorption/desorption properties were tested in both aqueous and organic dyes.

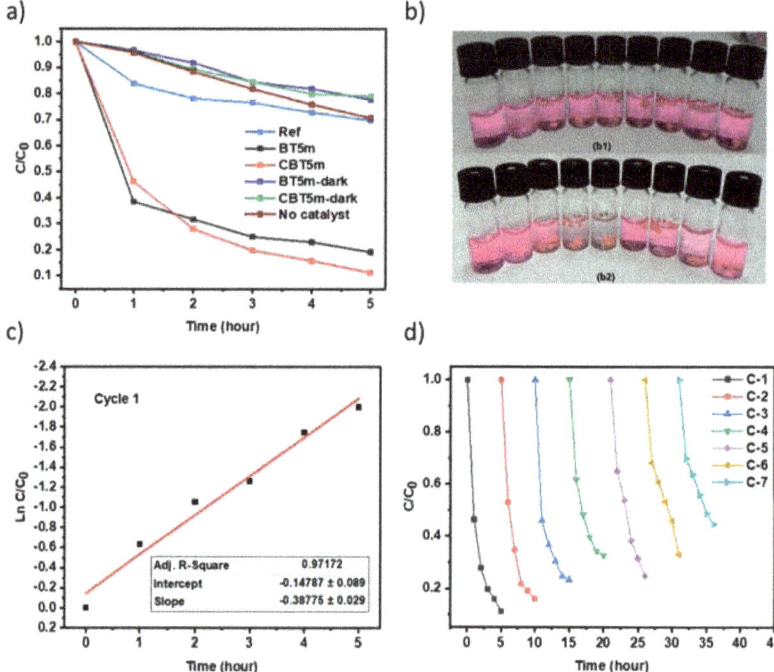

Figure 2. RhB dye degradation of BT5M, CBT5m, reference and RhB dye solution (no catalyst) under visible light irradiation (BT5m-dark and CBT5m-dark were not exposed to visible light) (**a**). Digital images of PS-co-DVB-derived beads in RhB solution, from left to right: RhB Dye, REF, B5m, BT5m, CBT5m, BT5s, B7m, BT7m, BT10m before and after 5 h visible light irradiation (**b**). Pseudo-first order kinetic fitting data of cycle 1 (pH:6.3, T = 25 °C, bead/RhB:0.055/0.0015 g·L^{-1}) (**c**). Cyclic photocatalytic RhB dye degradation with several run cycles in the presence of CBT5m under visible light irradiation (**d**) [7]. Copyright 2021, John Wiley & Sons.

Afterwards, g-C_3N_4-incorporated beads were subjected to a photo-induced surface modification by employing vinylsulfonic acid (VSA) and 4-vinyl pyridine (VP) under visible light irradiation. According to the results, VSA photo-grafting was analyzed successfully by elemental mapping via EDX along with elemental analysis via detected sulfur content. On the other hand, VP photo-grafting was confirmed via combustive elemental analysis together with FT-IR successfully.

In other words, beads donated with photoactivity via metal-free semiconductor g-C_3N_4 integration can be exploited for dye photodegradation and as an acid-base catalyst after post-functionalization through a simple photo-induced surface modification.

2.2. Synthesis of g-C_3N_4 Integrated Hydrogels and Subsequent Post-Modifications

As a particular class of hydrophilic polymers, hydrogels possess a high water content through their crosslinked porous networks. A facile synthesis of a hydrogel can be simply described as a combination of a water-soluble monomer and a multi-functional crosslinker in the presence of an initiator (mainly radical-based systems). In this context, firstly, g-C_3N_4 embedded hydrogel synthesis, thereafter its further subsequent visible light-assisted post modifications will be described (Scheme 2).

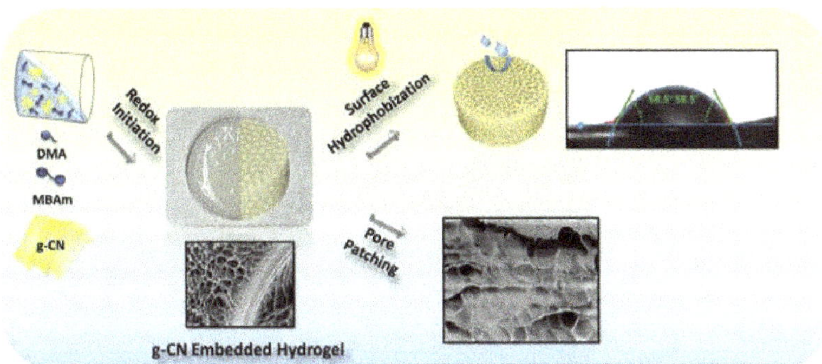

Scheme 2. Schematic overview of g-C$_3$N$_4$-embedded hydrogel fabrication and its subsequent photo-induced post-modifications [8]. Copyright 2021, Beilstein Journal of Organic Chemistry.

Template g-C$_3$N$_4$-embedded hydrogel was formed by free-radical polymerization in the presence of a redox couple (ascorbic acid-hydrogen peroxide), water-soluble monomer (*N,N*-Dimethylacrylamide, DMA) with a crosslinker (*N,N'*-methylenebisacrylamide, MBA), in an aqueous dispersion of g-C$_3$N$_4$. Gelation proceeded in 3 h and the resulting hydrogel further employed surface hydrophobization by being immersed into 4-methyl-5-vinylthiazole (vTA) under visible light irradiation. Drastic morphology change over the pore closure, observed via SEM investigation (Figure 3(a1,a2)), and sulfur detection via EDX measurement were supportive evidences of performed surface photo-modification (Figure 3b). Moreover, resulting hydrophobized hydrogel showed an enhanced water contact angle result (58.5° over 40 s), unlike the initial hydrophilic hydrogel (Figure 3c). In addition, the sustained cation release performance by investigating several cations in both samples was confirmed, based on controlled release after surface hydrophobization (Figure 3d).

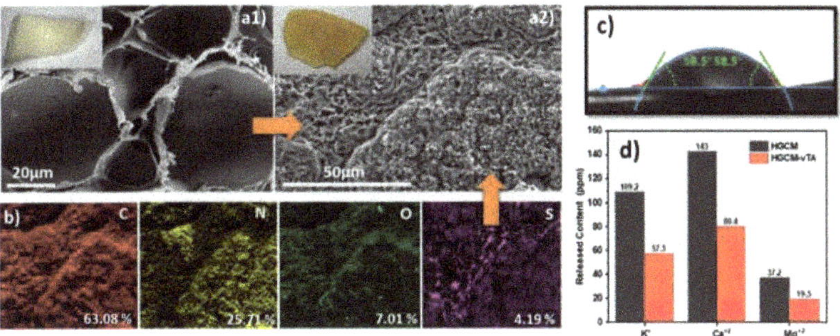

Figure 3. Scanning electron microscopy (SEM) images of HGCM (**a1**) and HGCM-vTA (**a2**) in combination with HGCM-vTA elemental mapping results via EDX (**b**). Water contact angle result of HGCM-vTA (**c**). Inductively coupled plasma optical emission spectroscopy (ICP-OES) results of released K$^+$, Ca^{2+} and Mg^{2+} content for HGCM and HGCM-vTA (**d**) [8]. Copyright 2021, Beilstein Journal of Organic Chemistry.

Considering the SEM results exhibited the highly porous structure of host hydrogel, pore substructuring was performed by taking advantage of the embedded g-C$_3$N$_4$ photoactivity. Regarding that, host hydrogel was swollen by various vinyl monomers and exposed to visible light. According to the SEM investigation, the closure of pores via polymer layer formation was observed, and the detection of functional groups via FT-IR underlined the efficient modification (Figure 4a–c).

3. Conclusions

The combination of a metal-free semiconductor g-C$_3$N$_4$ with polymer materials is highly exciting, and offers many advanced materials, as shown in the present contribution. Photoactivity of g-C$_3$N$_4$ can be harnessed to synthesize polymers through photoredox chemistry (either forming active radicals through charge transfer processes, excited electrons, or a positive hole can be directly employed), or alternatively g-C$_3$N$_4$ can be dispersed and embedded in polymer networks to grant photoactivity. Merging polymers with colloidal g-C$_3$N$_4$ materials is still at the infant stage; however, one can distill versatile synthetic-polymer chemistry knowledge with semiconductor science to design functional materials that can address soft materials, conducting polymers, hybrid polymer materials, agricultural delivery systems and catalysis.

Author Contributions: The authors contributed equally.

Funding: Max Planck Society.

Data Availability Statement: All the data produced are available from authors.

Acknowledgments: The authors thank the Max Planck Society for funding and providing the infrastructure. We thank Markus Antonietti for the continuous support and mentoring.

Conflicts of Interest: The authors declare no conflict of interest.

References

1. Zhou, Z.; Zhang, Y.; Shen, Y.; Liu, S.; Zhang, Y. Molecular engineering of polymeric carbon nitride: Advancing applications from photocatalysis to biosensing and more. *Chem. Soc. Rev.* **2018**, *47*, 2298–2321. [CrossRef] [PubMed]
2. Giusto, P.; Cruz, D.; Heil, T.; Arazoe, H.; Lova, P.; Aida, T.; Comoretto, D.; Patrini, M.; Antonietti, M. Shine Bright Like a Diamond: New Light on an Old Polymeric Semiconductor. *Adv. Mater.* **2020**, *32*, 1908140. [CrossRef] [PubMed]
3. Giusto, P.; Kumru, B.; Zhang, J.; Rothe, R.; Antonietti, M. Let a Hundred Polymers Bloom: Tunable Wetting of Photografted Polymer-Carbon Nitride Surfaces. *Chem. Mater.* **2020**, *32*, 7284–7291. [CrossRef]
4. Savateev, A.; Antonietti, M. Heterogeneous Organocatalysis for Photoredox Chemistry. *ACS Catal.* **2018**, *8*, 9790–9808. [CrossRef]
5. Kumru, B.; Antonietti, M. Colloidal properties of the metal-free semiconductor graphitic carbon nitride. *Adv. Colloid Interface Sci.* **2020**, *283*, 102229. [CrossRef] [PubMed]
6. Kumru, B.; Molinari, V.; Hilgart, M.; Rummel, F.; Schäffler, M.; Schmidt, B.V.K.J. Polymer grafted graphitic carbon nitrides as precursors for reinforced lubricant hydrogels. *Polym. Chem.* **2019**, *10*, 3647–3656. [CrossRef]
7. Esen, C.; Antonietti, M.; Kumru, B. Upgrading poly(styrene-co-divinylbenzene) beads: Incorporation of organomodified metal-free semiconductor graphitic carbon nitride through suspension photopolymerization to generate photoactive resins. *J. Appl. Polym. Sci.* **2021**, *138*, 50879. [CrossRef]
8. Esen, C.; Kumru, B. Photoinduced post-modification of graphitic carbon nitride-embedded hydrogels: Synthesis of 'hydrophobic hydrogels' and pore substructuring. *Beilstein J. Org. Chem.* **2021**, *17*, 1323–1334. [CrossRef] [PubMed]
9. Esen, C.; Antonietti, M.; Kumru, B. Oxidative Photopolymerization of 3,4-Ethylenedioxythiophene (EDOT) via Graphitic Carbon Nitride: A Modular Toolbox for Attaining PEDOT. *ChemPhotoChem* **2021**, *5*, 857–862. [CrossRef]

MDPI

Abstract

Vitrimerization of Poly(butylene succinate) By Reactive Melt Mixing Using Zn(II) Epoxy-Vitrimer Chemistry [†]

Christos Panagiotopoulos[iD]**, Dimitrios Korres and Stamatina Vouyiouka ***

Laboratory of Polymer Technology, School of Chemical Engineering, National Technical University of Athens, Zographou Campus, 157 80 Athens, Greece; chpanagiotopoulos@mail.ntua.gr (C.P.); dmkorres@central.ntua.gr (D.K.)

* Correspondence: mvuyiuka@central.ntua.gr

† Presented at the 2nd International Online Conference on Polymer Science—Polymers and Nanotechnology for Industry 4.0, 1–15 November 2021; Available online: https://iocps2021.sciforum.net/.

Abstract: Vitrimers constitute a new class of covalent adaptable networks (CANs), in which thermally stimulated associative exchange reactions allow the topological rearrangement of the dynamic network while keeping the number of bonds and the crosslink density constant. The current study proposed a solvent-free method to synthesize vitrimers by reactive melt mixing using a commercial biobased/biodegradable polyester, poly(butylene succinate), PBS. More specifically, a two-step process was followed; the first step involved reactive mixing of PBS with the crosslinker (diglycidyl ether of bisphenol A, DGEBA) and the transesterification catalyst (Zinc(II) acetylacetonate hydrate, Zn(acac)2) in a twin-screw mini-compounder, in order to incorporate the epoxy groups in the polymer backbone. The second step (vitrimerization) comprised a crosslinking process of the homogenous mixtures in a vacuum oven at 170 °C, resulting in the formation of a dynamic crosslinked network with epoxy moieties serving as the crosslinkers. By tuning the crosslinker content (0–10% mol with respect to PBS repeating unit) and the Zinc(II) catalyst to crosslinker ratio (0 to 1), tailor-made vitrimers were prepared with high insolubility and improved melt strength. Moreover, PBS vitrimers could still be reprocessed by compression molding after the crosslinking, which enables the recycling process. This work was made possible by the "Basic Research Programme, NTUA, PEVE 2020 NTUA" [PEVE0050] of the National Technical University of Athens and is gratefully acknowledged.

Keywords: poly(butylene succinate); epoxy-based vitrimers; polyesters; crosslinking; reprocessability; recycling

---

Citation: Panagiotopoulos, C.; Korres, D.; Vouyiouka, S. Vitrimerization of Poly(butylene succinate) By Reactive Melt Mixing Using Zn(II) Epoxy-Vitrimer Chemistry. *Mater. Proc.* **2021**, *7*, 10. https://doi.org/10.3390/IOCPS2021-11588

Academic Editor: Gianluca Cicala

Published: 5 November 2021

Publisher's Note: MDPI stays neutral with regard to jurisdictional claims in published maps and institutional affiliations.

Supplementary Materials: The following supporting information can be downloaded at: https://www.mdpi.com/article/10.3390/IOCPS2021-11588/s1.

Author Contributions: Conceptualization, S.V.; methodology, S.V. and C.P.; validation, C.P. and D.K.; investigation, C.P.; data curation, C.P. and D.K.; writing—original draft preparation, C.P.; writing—review and editing, S.V. and C.P.; visualization, C.P. and D.K.; supervision, S.V.; project administration, S.V. All authors have read and agreed to the published version of the manuscript.

Funding: This work was funded by the "Basic Research Programme, NTUA, PEVE 2020 NTUA" [PEVE0050] of the National Technical University of Athens and is gratefully acknowledged.

Institutional Review Board Statement: Not applicable.

Informed Consent Statement: Not applicable.

Data Availability Statement: The data presented in this study are available upon request from the corresponding author. The data are not publicly available due to restrictions regarding an on-going funded research.

Conflicts of Interest: The authors declare no conflict of interest.

---

materials
proceedings

Abstract

Evaluation of the Parameters of Poly(Butylene succinate) Enzymatic Polymerization [†]

Christina I. Gkountela, Dimitrios N. Markoulakis, Dimitrios M. Korres and Stamatina N. Vouyiouka *

Laboratory of Polymer Technology, School of Chemical Engineering, National Technical University of Athens, 15780 Athens, Greece; cgkountela@mail.ntua.gr (C.I.G.); dimark97@hotmail.com (D.N.M.); dmkorres@central.ntua.gr (D.M.K.)
* Correspondence: mvuyiuka@central.ntua.gr
† Presented at the 2nd International Online Conference on Polymer Science—Polymers and Nanotechnology for Industry 4.0, 1–15 November 2021; Available online: https://iocps2021.sciforum.net/.

Abstract: Poly(butylene succinate) (PBS) is a bio-based and biodegradable polyester that can be used in numerous applications, ranging from clothing to food packaging and from the car industry to the biomedical sector (e.g., drug release systems). The conventional polymerization method of PBS requires the presence of metal-based transesterification catalysts (e.g., titanium-based catalysts) and high reaction temperatures (T > 150 °C). However, under these conditions side reactions may occur along with undesirable yellowing. Green polymerization routes such as biocatalysis are being developed. However, there is a very limited literature on the enzymatic synthesis of PBS. Additionally, in most of the works where high-molecular-weight PBS is produced from the typical monomers (BDO and DES), several drawbacks, e.g., the use of various solvents for polymer isolation and the requirement of high vacuum for by-products removal may impede the process being scaled up. On that basis, an eco-friendly, solvent-free, enzyme-based process for the production of PBS was applied. It was conducted in two steps with the use of Novozym 435: the first at 40 °C, under atmospheric pressure for 24 h, and the second at 90 °C, 20 mbar for 2 h. This work focused on the optimization of the second step's conditions, by varying reaction temperature (80–95 °C), pressure (20 mbar, 200 mbar) and reaction time (2 h, 6 h). Based on the optimization results, the process was scaled up (ca. 10 g of product). A PBS grade free of thermal degradation and metal catalyst residues, of weight-average molecular weight 4700 g/mol and melting point 103 °C, was obtained.

Keywords: bio-based/biodegradable polyester; enzymatic polymerization; lipase; poly(butylene succinate)

Citation: Gkountela, C.I.; Markoulakis, D.N.; Korres, D.M.; Vouyiouka, S.N. Evaluation of the Parameters of Poly(Butylene succinate) Enzymatic Polymerization. *Mater. Proc.* **2021**, 7, 11. https://doi.org/10.3390/IOCPS2021-11274

Academic Editor: Marina Arrieta

Published: 1 November 2021

Publisher's Note: MDPI stays neutral with regard to jurisdictional claims in published maps and institutional affiliations.

Supplementary Materials: The poster presentation is available online at https://www.mdpi.com/article/10.3390/IOCPS2021-11274/s1.

Institutional Review Board Statement: Not applicable.

Informed Consent Statement: Not applicable.

Data Availability Statement: The data presented in this study are available on request from the corresponding author. The data are not publicly available due to time restrictions.

materials proceedings

Proceeding Paper

Optimization of Foamed Polyurethane/Ground Tire Rubber Composites Manufacturing [†]

Adam Olszewski, Paulina Kosmela, Łukasz Zedler, Krzysztof Formela and Aleksander Hejna *

Department of Polymer Technology, Gdańsk University of Technology, Narutowicza 11/12,
80-233 Gdańsk, Poland; adam.olszewski@student.pg.edu.pl (A.O.); paulina.kosmela@pg.edu.pl (P.K.);
lukasz.zedler@pg.edu.pl (Ł.Z.); krzform1@pg.edu.pl (K.F.)
* Correspondence: aleksander.hejna@pg.gda.pl
† Presented at the 2nd International Online Conference on Polymer Science—Polymers and Nanotechnology for Industry 4.0, 1–15 November 2021; Available online: https://iocps2021.sciforum.net/.

Abstract: The development of the automotive sector and the increasing number of vehicles all over the world poses multiple threats to the environment. One of them, probably not so emphasized as others, is the enormous amount of post-consumer car tires. Due to the potential fire threat, waste tires are considered as dangerous waste, which should not be landfilled, so it is essential to develop efficient methods of their utilization. One of the possibilities is their shredding and application of resulting ground tire rubber (GTR) as filler for polymer composites, which could take advantage of the excellent mechanical performance of car tires. Nevertheless, due to the poor compatibility with majority of polymer matrices, prior to the application, surface of GTR particles should be modified and activated. In the presented work, the introduction of thermo-mechanically modified GTR into flexible foamed polyurethane matrix was analyzed. Isocyanates can be found among the compounds applied during manufacturing of polyurethane foams, which are able to react and generate covalent bonds with the functional groups present on the surface of modified GTR. Such an effect can noticeably enhance the interfacial interactions and boost up the mechanical performance. Nevertheless, it requires the adjustment of formulations used during manufacturing of foams. Therefore, for better understanding of the process foams with varying isocyanate index (from 0.8 to 1.2) were prepared with and without taking into account the possible interactions with functional groups of GTR. For comparison, an unfilled matrix and composite containing deactivated GTR were also prepared.

Keywords: polyurethane foam; ground tire rubber; rubber modification; surface activation polymer composites

---

Citation: Olszewski, A.; Kosmela, P.; Zedler, Ł.; Formela, K.; Hejna, A. Optimization of Foamed Polyurethane/Ground Tire Rubber Composites Manufacturing. *Mater. Proc.* **2021**, 7, 12. https://doi.org/10.3390/IOCPS2021-11244

Academic Editor: Shin-ichi Yusa

Published: 30 October 2021

Publisher's Note: MDPI stays neutral with regard to jurisdictional claims in published maps and institutional affiliations.

1. Introduction

Polyurethane (PU) foams are commonly applied in various branches of the industry due to their wide range of easily adjustable applications [1]. Their properties strongly depend on the applied formulation, particularly on the ratio of isocyanate and hydroxyl groups introduced by PU components, expressed by the isocyanate index. Hydroxyl groups are present in the structure of polyols, the major components of polyurethanes. Moreover, they can be found in water applied as a blowing agent and in the chemical structure of fillers introduced into the PU matrix [2]. Hydroxyls are primarily present in organic, plant-based fillers such as cellulose, wood flour, or natural fibers [3]. However, they can also be found in materials previously subjected to oxidative conditions during different industrial processes [4]. Among such materials can be mentioned ground tire rubber (GTR) generated during the recycling of post-consumer car tires. During the shredding of tires, hydroxyl groups can be introduced onto the surface of GTR particles due to oxidation occurring during the reduction in particle size [5]. The incorporation of such materials affects the desired balance between isocyanate and hydroxyl groups in the PU formulation, which may influence performance of composite materials [6]. Członka et al. [7] reported that

---

solid waste from the leather industry, rich in hydroxyl groups, might noticeably affect the foaming kinetics and cellular structure of PU foams. The unfavorable influence of natural fillers on the performance of PU foams was reported by Zieleniewska et al. [8].

The effect of additional hydroxyl groups is particularly noticeable for foams, characterized by the lower values of the isocyanate index. For rigid foams with values of isocyanate index around 2.0 or higher, the effect is not so strong [9]. On the other hand, for flexible foams, even a tiny amount of filler may noticeably affect the balance between functional groups, as reported by Silva et al. [10] for incorporation of waste rubber particles. Despite incorporating filler showing superior mechanical performance, composite foams' strength was lower than the unfilled matrix. It points to the weakness of the PU network, attributed to its reduced cross-linking [11]. The effect may be particularly significant for the fillers very rich in hydroxyl groups, which was reported in our previous paper when GTR was additionally oxidized with potassium permanganate [12].

Therefore, to prepare PU composite foams efficiently, it is important to consider the influence of the chemical structure of fillers and the presence of functional groups, which may affect the overall isocyanate index. The presented work aimed to investigate the effect of GTR hydroxyl groups on the mechanical performance of flexible PU foams prepared with varying isocyanate index. The tensile and compression tests were performed to assess the foam modifications' impact and describe it qualitatively and quantitatively.

2. Materials and Methods

2.1. Materials

Table 1 presents the materials applied in the presented work.

Table 1. The list of materials used in the presented work.

Material	Origin	Details
PU foams preparation		
Rokopol® F3000	PCC Group (Brzeg Dolny, Poland)	Polyol, hydroxyl value (L_{OH}) = 53–59 mg KOH/g
Rokopol® V700		Polyol, L_{OH} = 225–250 mg KOH/g
Glycerol	Sigma Aldrich (Poznań, Poland)	L_{OH} = 1800 mg KOH/g
SPECFLEX NF 434	M. B. Market Ltd. (Baniocha, Poland)	Polymeric methylenediphenyl-4,4′-diisocyanate, free -NCO content = 29.5%
PC CAT® TKA30	Performance Chemicals (Belvedere, UK)	Catalyst, potassium acetate
Dabco33LV	Air Products (Allentown, PA, USA)	Catalyst, 1,4-diazabicyclo [2.2.2]octane in dipropylene glycol, 3 wt% solution
Dibutyltin dilaurate	Sigma Aldrich (Poznań, Poland)	Catalyst, organic tin compound
Distilled water	-	Chemical blowing agent
Ground tire rubber	Recykl S.A. (Śrem, Poland)	Filler, mean particle size = 0.6 mm, L_{OH} = 61.7 ± 3.0 mg KOH/g
Deactivation of GTR -OH groups		
Acetone		Solvent
Toluene diisocyanate		Free -NCO content = 42%
Dibutylamine	Sigma Aldrich (Poznań, Poland)	Analyte solution
Chlorobenzene		Solvent
Hydrochloric acid		Titrant
3′,3″,5′,5″-Tetrabromophenol-sulfonphthalein		Indicator

2.2. Deactivation of Ground Tire Rubber Hydroxyl Groups

To prepare GTR with deactivated hydroxyl groups, the equivalent amounts of GTR and TDI were placed in a glass flask in acetone. Components were mechanically mixed and left for 24 h in a dark place at room temperature. Then, GTR was taken out, washed with acetone to remove the excess of TDI, and dried to remove the solvent. To ensure the successful deactivation of hydroxyl groups, the hydroxyl value of GTR was determined using the modified test method for isocyanate groups, as described in our other work [13].

2.3. Preparation of Polyurethane/Ground Tire Rubber Composite Foams

A single-step method was applied to manufacture PU/GTR composite foams on a laboratory scale. The predetermined amount of GTR particles and polyols were mechanically mixed (1000 rpm, 60 s) to properly disperse filler particles in polyol mixture. Then, catalysts, blowing agent, and isocyanate were introduced, and the mixture was mechanically mixed (1800 rpm, 10 s) and poured into a closed aluminum mold with dimensions of $20 \times 10 \times 4$ cm^3. Prior to the structure and performance analysis, foams were conditioned at room temperature and 60% average humidity for 24 h.

As mentioned above, for a better understanding of the interactions between isocyanates and functional groups of GTR filler, foams with varying isocyanate index (from 0.8 to 1.2) were prepared. Except for neat foams without the GTR addition (named PU$_X$), three series of composite foams were prepared, which were named D-GTR$_X$, N-GTR$_X$, and C-GTR$_X$, where D indicated deactivated GTR, N not considered, and C considered in the isocyanate index calculation. For all samples, X indicates the isocyanate index applied in the formulation. Table 2 shows the formulations of prepared composite foams. All foams were characterized by a similar level of apparent density—205 $\pm$ 6 kg/m^3.

Table 2. Formulations of prepared PU and PU/GTR foams.

Component	Neat Foam							GTR Deactivated/Not Considered							GTR Considered						
	Content, wt%																				
F3000	35.4	34.2	33.7	33.2	32.7	32.2	31.2	29.5	28.5	28.1	27.6	27.2	26.8	26.0	28.3	27.3	26.8	26.4	25.9	25.5	24.6
V700	35.4	34.2	33.7	33.2	32.7	32.2	31.2	29.5	28.5	28.1	27.6	27.2	26.8	26.0	28.3	27.3	26.8	26.4	25.9	25.5	24.6
Glycerol	0.9	0.8	0.8	0.8	0.8	0.8	0.8	0.7	0.7	0.7	0.7	0.7	0.6	0.6	0.7	0.7	0.6	0.6	0.6	0.6	0.6
DBTDL	0.6	0.6	0.6	0.6	0.6	0.6	0.6	0.5	0.5	0.5	0.5	0.5	0.5	0.5	0.5	0.5	0.5	0.5	0.5	0.5	0.4
33LV	0.4	0.4	0.4	0.4	0.4	0.4	0.4	0.4	0.3	0.3	0.3	0.3	0.3	0.3	0.3	0.3	0.3	0.3	0.3	0.3	0.3
TKA30	0.4	0.4	0.4	0.4	0.4	0.4	0.4	0.4	0.3	0.3	0.3	0.3	0.3	0.3	0.3	0.3	0.3	0.3	0.3	0.3	0.3
Water	0.4	0.4	0.3	0.3	0.3	0.3	0.3	0.3	0.3	0.3	0.3	0.3	0.3	0.3	0.3	0.3	0.3	0.3	0.3	0.3	0.3
pMDI	26.6	28.9	30.0	31.1	32.2	33.2	35.2	22.1	24.1	25.0	25.9	26.8	27.7	29.3	24.5	26.6	27.7	28.7	29.6	30.6	32.3
GTR/modified GTR	-	-	-	-	-	-	-	16.7	16.7	16.7	16.7	16.7	16.7	16.7	16.7	16.7	16.7	16.7	16.7	16.7	16.7
Isocyanate:hydroxyl ratio	0.8	0.9	0.95	1.0	1.05	1.1	1.2	0.8	0.9	0.95	1.0	1.05	1.1	1.2	0.8	0.9	0.95	1.0	1.05	1.1	1.2

2.4. Characterization Techniques

The tensile performance of prepared foams was investigated following ISO 1798 standard using Zwick/Roell Z020 tensile tester (Ulm, Germany). The beam-shaped samples were used. Their dimensions, measured using slide caliper with 0.1 mm accuracy, were $10 \times 10 \times 100$ mm^3. During tensile tests, constant tension rate of 500 mm/min was applied.

The compressive performance of PU foams was investigated following ISO 604 standard using Zwick/Roell Z020 tensile tester (Ulm, Germany). The cylindric samples were used. Their dimensions (height and diameter), measured using slide caliper with 0.1 mm accuracy, were 20 mm $\times$ 20 mm. During tests, a constant compression rate of 15%/min was applied. Foams were tested until reaching 60% deformation.

3. Results and Discussion

Figure 1 presents the impact of the isocyanate index on the tensile strength of prepared foams. It can be seen that the mechanical performance significantly depends on the isocyanate index applied during the preparation of foams. It is associated with the development of a polyurethane network during reactions between hydroxyl and isocyanate groups. Typically, increasing the isocyanate index is beneficial for the tensile performance of flexible polyurethane foams. Such an effect was noted by Prociak et al. [14] and Lee et al. [15]. In the presented case, the tensile strength was almost directly proportional to the isocyanate index. For higher values of isocyanate index, especially exceeding 1.0, the excess of isocyanate may react with hydroxyl groups present on the surface of filler and strengthen the interface.

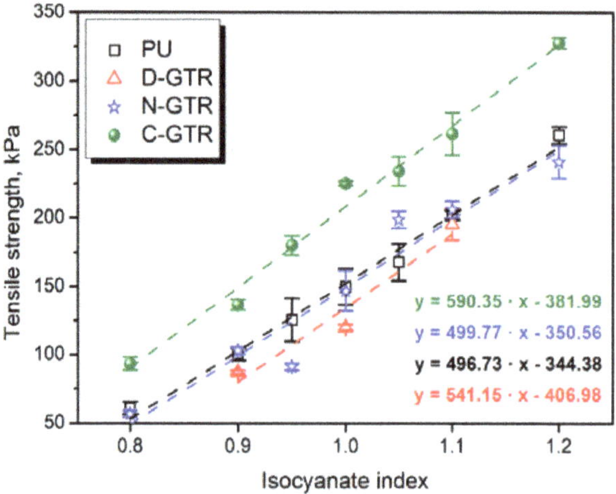

Figure 1. The impact of isocyanate index on the tensile strength of prepared PU foams.

Given the influence of applied formulation, a significant difference was observed between samples which formulation considered or not considered the hydroxyl values of GTR. When applied filler was not taken into account during isocyanate index calculation, the tensile strength of composites was slightly lower than the neat matrix. It suggests that for N-GTR foams, rubber particles were at least to some extent bonded with polyurethane matrix. However, it also indicates that the interfacial interactions between matrix and filler were relatively poor because the tensile strength of GTR itself is in the range of MPas [16]. At the same time, the worst results were noted when deactivated GTR particles were introduced. It indicates that when the inert filler is incorporated into the polyurethane matrix, its development during foaming is affected, and the resulting strength of the final composite is reduced [17].

Significantly, the best results were observed for C-GTR foams, attributed to the higher amount of isocyanate used. Compared to N-GTR foams, considering the hydroxyl value of GTR filler in calculation of isocyanate index required ~10% more isocyanate. As a result, more covalent linkages in the polyurethane network were developed, which strengthened the material. Figure 2 shows plots aimed to provide more quantitative information about the amount of GTR hydroxyls reacted with isocyanate. It shows the dependence between applied and calculated isocyanate index assuming different reactivity of GTR functional groups. Full reactivity means that all hydroxyl groups present on the surface of GTR particles reacted with isocyanates during foaming. Therefore, applied and calculated values of the isocyanate index are equal. No reactivity means that the isocyanates theoretically reacted only with polyols and water. In such a case, the actual values of the isocyanate index (calculated ones) would be substantially higher, as more isocyanate was introduced into formulations. Experimental data points are obtained by applying tensile strength-isocyanate index dependence for PU samples presented in Figure 1 (y = 496.73x − 344.38) to analyze the C-GTR samples. All data points lie between full reactivity and no reactivity, indicating that only part of GTR hydroxyls involved reactions with isocyanates. It was affected mainly by the lower reactivity of these groups compared to polyols' hydroxyls and steric hindrance caused by bulk GTR particles [18]. Presented data indicate that from 7 to 51% of GTR hydroxyls reacted with isocyanates depending on the sample. For most samples, the average value was 28 ± 5%.

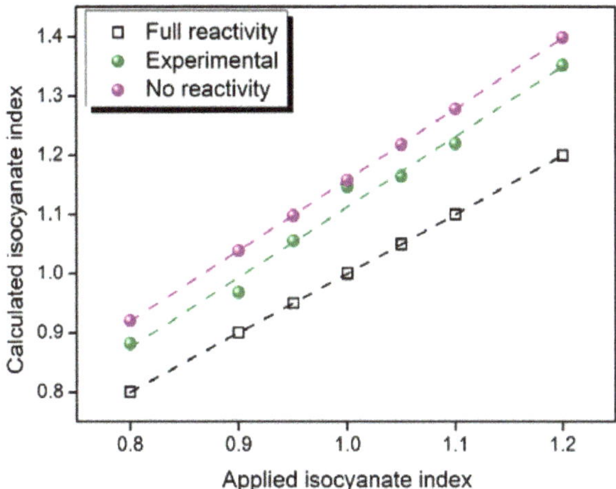

Figure 2. Calculated and applied values of isocyanate index based on the tensile tests and used formulations.

Figure 3 presents the compressive performance of prepared foams depending on the isocyanate index. The increase in compressive strength was associated with better development of polyurethane network and higher cross-link density of a material. Such a strengthening effect was observed by other researchers [14,19]. Presented data also indicate that functional groups of GTR affect the performance of the PU matrix. For N-GTR samples, strength deterioration was noted, despite the superior performance of filler compared to polyurethane.

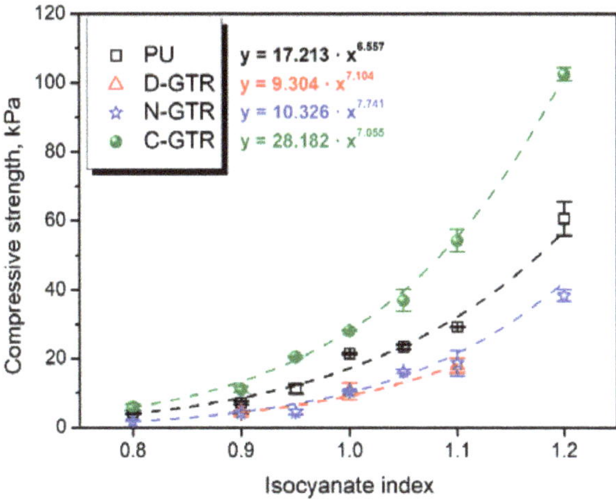

Figure 3. Compressive strength of prepared foams depending on the isocyanate index.

Figure 4, similar to Figure 2, presents the quantitative information about the interaction of GTR particles with the PU matrix. The experimental data points for C-GTR samples were calculated from the compressive strength-isocyanate index curve described by the power equation ($y = 17.213 \times x^{6.557}$). Similar to tensile-based dependence, all data points indicate partial reactivity between isocyanate groups and hydroxyls on the GTR surface. However, compression data indicates higher reactivity in the range of 43–74% for most samples 48–57%. Around twice as high reactivity compared to the tensile-based dependence is attributed to the different deformation mechanisms and GTR particles' contribution. During tension, the cohesion of material is crucial so that it can withstand deformation. Therefore, heterogeneity and insufficient interfacial adhesion result in discontinuity of material and reduced tensile strength. The filler itself may enhance the strength of the composite. However, such an effect is usually noted for fibrous fillers, which can transfer stress during tension rather than particulate fillers [20]. Considering compression, the force is acting on the material in the opposite direction. Hence, the impact of filler is different. For an efficient reinforcement, the filler itself should withstand high compressive forces, and its impact on the PU foams' cellular structure is essential [4,6,12]. Due to GTR particles' characteristics (shape, aspect ratio, and mechanical performance), their impact on the foams' compressive performance was more substantial than for tensile strength. Therefore, compression-based dependence suggested a greater extent of GTR reaction with isocyanates compared to tension.

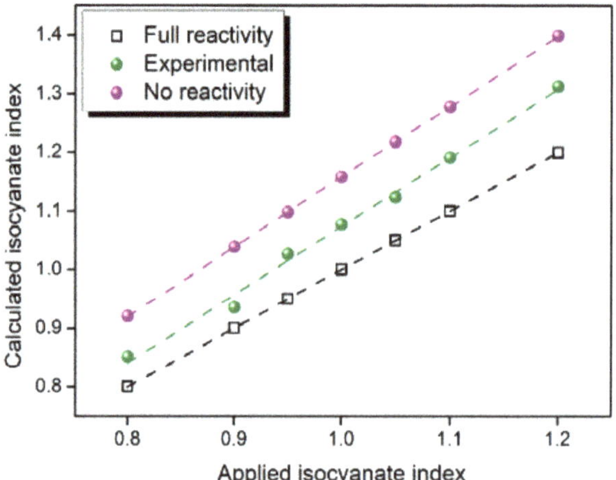

Figure 4. Calculated and applied values of isocyanate index based on the compression tests and used formulations.

4. Conclusions

The presented research aimed to investigate the impact of hydroxyl groups present on the surface of ground tire rubber particles on balance between the isocyanate and hydroxyl groups in the polyurethane system applied to produce flexible PU composite foams. Tensile-based dependences indicated that around 23–33% of GTR hydroxyls reacted with isocyanates, while compression tests suggested higher values in the range of 48–57%. The differences were associated with the filler performance and different modes of deformation. Nevertheless, despite the lack of chemical analysis, mechanical tests and calculations of foams' formulations pointed to partial reactivity of applied filler with isocyanates, which should be considered, especially during the development of flexible PU composite foam-based products on an industrial scale.

Author Contributions: Conceptualization, P.K. and A.H.; methodology, A.O. and P.K.; validation, P.K. and A.H.; formal analysis, A.H.; investigation, A.O. and P.K.; resources, A.H.; data curation, A.O.; writing—original draft preparation, A.H.; writing—review and editing, A.O., P.K., Ł.Z. and K.F.; visualization, A.H.; supervision, A.H.; project administration, A.H.; funding acquisition, A.H. All authors have read and agreed to the published version of the manuscript.

Funding: This work was supported by The National Centre for Research and Development (NCBR, Poland) in the frame of LIDER/3/0013/L-10/18/NCBR/2019 project—Development of technology for the manufacturing of foamed polyurethane-rubber composites for the use as damping materials.

Institutional Review Board Statement: Not applicable.

Informed Consent Statement: Not applicable.

Data Availability Statement: Data is contained within the article. The data presented in this study are available in Optimization of foamed polyurethane/ground tire rubber composites manufacturing.

Conflicts of Interest: The authors declare no conflict of interest.

References

1. Hejna, A.; Kosmela, P.; Kirpluks, M.; Cabulis, U.; Klein, M.; Haponiuk, J.; Piszczyk, Ł. Structure, Mechanical, Thermal and Fire Behavior Assessments of Environmentally Friendly Crude Glycerol-Based Rigid Polyisocyanurate Foams. *J. Polym. Environ.* **2017**, *26*, 1854–1868. [CrossRef]
2. Gosz, K.; Kosmela, P.; Hejna, A.; Gajowiec, G.; Piszczyk, Ł. Biopolyols obtained via microwave-assisted liquefaction of lignin: Structure, rheological, physical and thermal properties. *Wood Sci. Technol.* **2018**, *52*, 599–617. [CrossRef]
3. Hejna, A.; Barczewski, M.; Andrzejewski, J.; Kosmela, P.; Piasecki, A.; Szostak, M.; Kuang, T. Rotational Molding of Linear Low-Density Polyethylene Composites Filled with Wheat Bran. *Polymers* **2020**, *12*, 1004. [CrossRef] [PubMed]
4. Kosmela, P.; Olszewski, A.; Zedler, Ł.; Burger, P.; Formela, K.; Hejna, A. Structural Changes and Their Implications in Foamed Flexible Polyurethane Composites Filled with Rapeseed Oil-Treated Ground Tire Rubber. *J. Compos. Sci.* **2021**, *5*, 90. [CrossRef]
5. Hejna, A.; Korol, J.; Przybysz-Romatowska, M.; Zedler, Ł.; Chmielnicki, B.; Formela, K. Waste tire rubber as low-cost and environmentally-friendly modifier in thermoset polymers—A review. *Waste Manag.* **2020**, *108*, 106–118. [CrossRef] [PubMed]
6. Piszczyk, Ł.; Hejna, A.; Formela, K.; Danowska, M.; Strankowski, M. Effect of ground tire rubber on structural, mechanical and thermal properties of flexible polyurethane foams. *Iran. Polym. J.* **2015**, *24*, 75–84. [CrossRef]
7. Członka, S.; Bertino, M.F.; Strzelec, K.; Strąkowska, A.; Masłowski, M. Rigid polyurethane foams reinforced with solid waste generated in leather industry. *Polym. Test.* **2018**, *69*, 225–237. [CrossRef]
8. Zieleniewska, M.; Przyjemska, K.; Chojnacki, P.; Ryszkowska, J. Modification of flexible polyurethane foams by the addition of natural origin fillers. *Polym. Degrad. Stabil.* **2016**, *132*, 32–40. [CrossRef]
9. Piszczyk, Ł.; Hejna, A.; Danowska, M.; Strankowski, M.; Formela, K. Polyurethane/ground tire rubber composite foams based on polyglycerol: Processing, mechanical and thermal properties. *J. Reinf. Plast. Compos.* **2015**, *34*, 708–717. [CrossRef]
10. Silva, N.G.S.; Cortat, L.I.C.O.; Orlando, D.; Mulinari, D.R. Evaluation of rubber powder waste as reinforcement of the polyurethane derived from castor oil. *Waste Manag.* **2020**, *116*, 131–139. [CrossRef] [PubMed]
11. Hejna, A.; Haponiuk, J.; Piszczyk, Ł.; Klein, M.; Formela, K. Performance properties of rigid polyurethane-polyisocyanurate/brewers' spent grain foamed composites as function of isocyanate index. *e-Polymers* **2017**, *17*, 427–437. [CrossRef]
12. Hejna, A.; Olszewski, A.; Zedler, Ł.; Kosmela, P.; Formela, K. The Impact of Ground Tire Rubber Oxidation with H_2O_2 and KMnO4 on the Structure and Performance of Flexible Polyurethane/Ground Tire Rubber Composite Foams. *Materials* **2021**, *14*, 499. [CrossRef] [PubMed]
13. Zedler, Ł.; Kosmela, P.; Olszewski, A.; Burger, P.; Formela, K.; Hejna, A. Recycling of Waste Rubber by Thermo-Mechanical Treatment in a Twin-Screw Extruder. *Proceedings* **2021**, *69*, 10. [CrossRef]
14. Prociak, A.; Malewska, E.; Bąk, S. Influence of Isocyanate Index on Selected Properties of Flexible Polyurethane Foams Modified with Various Bio-Components. *J. Renew. Mater.* **2016**, *4*, 78–85. [CrossRef]
15. Lee, C.S.; Ooi, T.L.; Chuah, C.H.; Ahmad, S. Effect of isocyanate index on physical properties of flexible polyurethane foams. *Malays. J. Sci.* **2007**, *26*, 91–98.
16. Zedler, Ł.; Kowalkowska-Zedler, D.; Colom, X.; Cañavate, J.; Saeb, M.R.; Formela, K. Reactive Sintering of Ground Tire Rubber (GTR) Modified by a Trans-Polyoctenamer Rubber and Curing Additives. *Polymers* **2020**, *12*, 3018. [CrossRef] [PubMed]
17. Hejna, A.; Kopczyńska, M.; Kozłowska, U.; Klein, M.; Kosmela, P.; Piszczyk, Ł. Foamed Polyurethane Composites with Different Types of Ash—Morphological, Mechanical and Thermal Behavior Assessments. *Cell. Polym.* **2016**, *35*, 287–308. [CrossRef]
18. Vilar, W.D. *Química e Tecnologia dos Poliuretanos*, 2nd ed.; Vilar Consultoria Técnica Ltd.: Rio de Janeiro, Brazil, 1998.
19. Abdel Hakim, A.A.; Nassar, M.; Emam, A.; Sultan, M. Preparation and characterization of rigid polyurethane foam prepared from sugar-cane bagasse polyol. *Mater. Chem. Phys.* **2011**, *129*, 301–307. [CrossRef]
20. Olcay, H.; Kocak, E.D. The mechanical, thermal and sound absorption properties of flexible polyurethane foam composites reinforced with artichoke stem waste fibers. *J. Ind. Text.* **2020**, 152808372093419. [CrossRef]

materials proceedings

Proceeding Paper

Gel Properties of Carboxymethyl Hyaluronic Acid/Polyacrylic Acid Hydrogels Prepared by Electron Beam Irradiation †

Alvin Kier R. Gallardo *, Lorna S. Relleve * and Alyan P. Silos

Chemistry Research Section, Philippine Nuclear Research Institute, Department of Science and Technology, Quezon City 1101, Philippines; apsilos@pnri.dost.gov.ph
* Correspondence: akrgallardo@pnri.dost.gov.ph (A.K.R.G.); lsrelleve@pnri.dost.gov.ph (L.S.R.)
† Presented at the 2nd International Online Conference on Polymer Science—Polymers and Nanotechnology for Industry 4.0, 1–15 November 2021; Available online: https://iocps2021.sciforum.net/.

Abstract: Semi-synthetic hydrogels made of carboxymethyl hyaluronic acid (CMHA) and polyacrylic acid (PAA) were synthesized using electron beam irradiation. CMHA, with a degree of substitution of 0.87 and a molecular weight of 149 kDa, was mixed with linear PAA and slightly crosslinked PAA (Carbopol). The equal weight ratio of the CMHA-Carbopol blends (10% CMHA, 10% Carbopol) was successfully crosslinked, even at a low irradiation dose of 20 kGy, producing a hydrogel with 60% gel fraction and 430 g/g degrees of swelling. The gel properties of this formulation showed good stability when exposed in PBS (pH 7.4) at 37 °C. Furthermore, the FT-IR spectra of the 10% CMHA, 10% Carbopol blends showed an increase in peak intensity at 1405 cm^{-1} due to the neutralization reaction between the COOH and COO$^-$ groups of PAA and CMHA polymers. The interaction effects between the concentration of CMHA and PAA and varying irradiation doses in the gel properties in CMHA-PAA hydrogels will be explored in a future study. Radiation crosslinking of biocompatible CMHA to other synthetic polymers, such as PAA, provides a cleaner method of producing biomaterials with tunable properties that are ideal for pharmaceutical, medical, and cosmetic applications.

Keywords: carboxymethyl hyaluronic acid; polyacrylic acid; radiation crosslinking; hydrogel

Citation: Gallardo, A.K.R.; Relleve, L.S.; Silos, A.P. Gel Properties of Carboxymethyl Hyaluronic Acid/Polyacrylic Acid Hydrogels Prepared by Electron Beam Irradiation. *Mater. Proc.* **2021**, *7*, 13. https://doi.org/10.3390/IOCPS2021-11220

Academic Editor: Shin-ichi Yusa

Published: 25 October 2021

Publisher's Note: MDPI stays neutral with regard to jurisdictional claims in published maps and institutional affiliations.

1. Introduction

Semi-synthetic hydrogels are widely used biomaterials, which have been extensively studied for use in pharmaceutical, medical, and cosmetic applications. The benefits of mixing synthetic and natural polymers allow semi-synthetic hydrogels to be tailored for specific applications [1]. Rosiak et al. (1999) reported that one of the most efficient ways of producing semi-synthetic gels for biomedical applications is through radiation crosslinking [2]. Previously, we successfully produced radiation-crosslinked biocompatible hydrogels made from pure carboxymethyl hyaluronic acid (CMHA) [3,4]. Despite successful crosslinking of pure CMHA, it is worthwhile to explore semi-synthetic hydrogels using CMHA combined with other synthetic polymers.

Polyacrylic acid (PAA) and its derivatives belong to the most commonly used synthetic water-soluble polymers applied to the production of hydrogels. Because PAA hydrogels contain ionizable carboxylic moieties, their gel properties, such as the degree of swelling, are greatly affected by the degree of neutralization as well as the pH and the ionic strength of the dispersing medium [5,6]. As a biomaterial, it demonstrated good biocompatibility and bioadhesive properties [7,8]. However, using it alone has disadvantages when administered biomedically, and these issues are resolved through combination with other polymers.

In this study, CMHA was incorporated with linear and slightly crosslinked PAA to produce semi-synthetic hydrogels through the use of electron beam irradiation. The gel properties, such as the gel fraction and the degree of swelling, were evaluated. The FT-IR spectra and thermal properties, as well as the in-vitro biodegradability of the CMHA-PAA hydrogels, were evaluated.

2. Materials and Methods

2.1. Materials

Cosmetic-grade and 1800 kDa HA (Stanford Chemicals Company, Lake Forest, CA, USA) sodium hydroxide (99%, RCI Labscan, Bangkok, Thailand), chloroacetic acid (99.5%, LOBA Chemie, Mumbai, India), and isopropyl alcohol (ACS reagent, J.T. Baker, Radnor, PA, USA) were all used as received for the carboxymethylation process. Analytical-grade sodium chloride (AR, Univar, Downers Grove, IL, USA) was used to prepare the mobile phase for the GPC analysis. Linear PAA and cosmetic-grade Carbopol-940 were obtained from Merck Sigma-Aldrich and Alysons' Chemical Enterprises, Inc. (Quezon City, Philippines), respectively. As with the in-vitro biodegradation experiment, a Phosphate Buffer Saline (PBS) tablet (Merck Sigma-Aldrich) was used.

2.2. Carboxymethylation of HA

Carboxymethyl hyaluronic acid was prepared using the previously reported protocol with a minor modification [4]. HA (20 g) was mixed with 50 mL of 45% *w/v* NaOH to form a paste and then mixed for 10 min. Isopropanol (2.0 L) was added and stirred for 1 h. About 60 g of solid chloroacetic acid was added and continuously stirred for 2.5 h. CMHA was collected by filtering the mixture through a non-woven fabric and then washed 3 times with isopropyl alcohol before being dissolved in 1.0 L deionized water. The CMHA solution was acidified until pH 7 with 12 N HCl and dialyzed using MWCO 12–14 kDa tubing for 3 days with frequent water changes. The solution was lyophilized, and the purified CMHA solid was stored at 4 °C for analysis and the preparation of hydrogels.

2.3. Determination of Molecular Weight and the Degree of Substitution

Potentiometric back titration was used to determine the degree of substitution (DS) of purified CMHA [4]. Briefly, cationic exchange resin (Amberlite H-form) was mixed in 5 mg/mL HA and 10 mg/mL CMHA aqueous solutions. The mixtures were swirled in a water bath for 2 h at 37 °C and then filtered through a stainless steel 200-mesh wire screen. The filtrates were passed through 0.45 μm and 0.22 μm syringe filters and then freeze-dried. About 0.1 g of dry acidified HA or CMHA was dissolved in 10 mL of standardized 0.1 M NaOH and diluted with 15 mL deionized water. This solution was then titrated with standardized 0.05 N HCl using phenolphthalein as an indicator. Blank titrations contained no sample. The DS was calculated based on the following equations:

$$n_{COOH} = \frac{(V_b - V)_{HCl}}{1000} \times C_{HCl} \tag{1}$$

$$A = \frac{n_{COOH\ (CMHA)}}{m_{CMHA}} - \frac{n_{COOH\ (HA)}}{m_{HA}} \tag{2}$$

$$DS = \frac{A \times MW_{DSU}}{1 - (A \times MW_1)} \tag{3}$$

where n_{COOH} = the mole of the carboxyl groups, V_b = the volume of HCl used to titrate the blanks, V = the volume of HCl used to titrate the sample, C_{HCl} = the concentration of HCl, MW_{DSU} = the molecular weight of the unsubstituted disaccharide unit, m = the mass of dry CMHA or HA, and MW_1 = the molecular weight of the carboxymethyl group.

Gel permeation chromatography (Shimadzu Prominence), equipped with a refractive index detector, was used to determine the average molecular weight (MW) of the purified CMHA. The chromatography columns used were the Tosoh TSK gel guard column PWXL and four TSK gel, serially connected, analytical columns (G6000 PWXL, G4000 PWXL, G3000 PWXL, and G2500 PWXL). Elution was performed using 0.2 M NaCl at a flow rate of 0.5 mL/min while the temperature of both the detector and columns was set to 40 °C. A calibration curve was constructed using polyethylene oxide (PEO) and polyethylene glycol (PEG) as standards. Molecular weights reported in this study are based on standards and are not absolute.

2.4. Preparation and Irradiation of Polymer Blends

Polymer blends of CMHA and PAA were prepared by combining the freeze-dried CMHA with linear PAA (CMHA-PAA) or with slightly crosslinked PAA (CMHA-Carbopol). The CMHA-PAA and CMHA-Carbopol blends were thoroughly mixed with water before being kneaded in plastic PE sheets and were allowed to stand overnight to allow for complete dissolution. Afterward, the samples were kneaded again to ensure the homogenization of the polymer blends prior to vacuum-sealing. Electron beam irradiation was carried out using a 2.5 MeV accelerator with a 12 mA current and a conveyor speed of 4.3–8.5 m/min. The resulting radiation-crosslinked hydrogels were then freeze-dried. The formulation of the samples is listed in Table 1.

Table 1. Formulation and irradiation doses of CMHA-PAA and CMHA-Carbopol blends.

Sample	CMHA Blend Concentration	Weight Ratio	Dose (kGy)
CMHA-PAA	37.5% CMHA, 2.5% PAA	15:1	20, 40
	35% CMHA, 5% PAA	7:1	20, 40
CMHA-Carbopol	10% CMHA, 10% Carbopol	1:1	20, 60, 120
Carbopol	10% Carbopol	-	20, 60, 120

2.5. Determination of Gel Properties

The gel fraction and the degree of swelling of the hydrogel samples were determined based on our previous study [3]. Approximately 0.2 g of the freeze-dried samples was immersed in 1.0 L of deionized water at room temperature for 72 h. Every 24 h, swollen samples were weighed and re-immersed in freshly changed deionized water. Swollen samples were then dried at 50 °C to a constant weight to obtain the gel fraction. The gel fraction and the degree of swelling were calculated as follows:

$$\text{Gel Fraction } (\%) = \frac{W_d}{W_i} \times 100\% \tag{4}$$

$$\text{Degree of Swelling } \left(\frac{g}{g}\right) = \frac{W_s - W_d}{W_d} \tag{5}$$

where W_d = the weight of the dried insoluble part after immersion for 3 days, W_i = the initial weight of the dried sample after irradiation, and W_s = the weight of the swollen gel. All measurements were completed in triplicates.

2.6. FT-IR and Thermogravimetric Analysis

The FT-IR spectra (600–4000 cm^{-1}) of freeze-dried hydrogel samples without sol fraction were obtained using a PerkinElmer Spectrum One FTIR spectrophotometer in attenuated total reflectance (ATR) mode. TGA thermograms were collected using a Netzsch STA449 F3 Jupiter at temperatures ranging from 25 to 1000 °C and at a heating rate of 10 °C/min in a nitrogen atmosphere.

2.7. In-Vitro Biodegradability

About 20.0 mg of freeze-dried hydrogels were placed in 20-mL vials and submerged in 4.0 mL of PBS (pH 7.4) or a 1:200 ratio of dried hydrogel to PBS. Samples were allowed to swell, were incubated at 37 °C, and were weighed periodically (at 1, 3, and 5 days). The pH of the buffer was adjusted to 7.4 (1.0 mM HCl and NaOH solution) and replenished as needed. The degree of swelling in the PBS solution was calculated from the weight of the swollen gel and the initial weight of the freeze-dried sample, as shown in the equation:

$$\text{Degree of Swelling in PBS} \left(\frac{g}{g} \right) = \frac{W_{s,PBS} - W_i}{W_i} \tag{6}$$

where $W_{s,PBS}$ = the weight of the swollen hydrogel in the PBS solution, while W_i = the weight of the freeze-dried hydrogel with insoluble fractions.

The in-vitro degradation of the hydrogel blends was evaluated by determining the weight loss of the freeze-dried hydrogel exposed in the PBS solution (pH 7.4) at 37 °C. Submerged hydrogels, in a 1:200 ratio of dried hydrogel to PBS, were collected periodically at intervals of 1, 3, and 5 days. Swollen samples were washed with PBS and then freeze-dried. The degradation was calculated as follows:

$$\text{Degradation } (\%) = \frac{W_{d,PBS}}{W_i} \times 100 \tag{7}$$

where $W_{d,PBS}$ = the weight of the freeze-dried CMHA submerged in PBS at day 1, 3, or 5.

3. Results and Discussion

3.1. Degree of Substitution and Molecular Weight of Carboxymethyl Hyaluronic Acid

The carboxymethylation of hyaluronic acid introduces a new functional group, which offers many possible prospects for the synthesis of new products from this simple polysaccharide with interesting chemical and biological properties. In radiation processing, the chemistry of carboxymethyl groups leads to the formation of radical intermediates, which are helpful in forming chemical bonds with other polymer chains, resulting in high-purity hydrogels. We previously attempted to make pure carboxymethyl hyaluronic hydrogels using high-energy radiation and discovered that higher DS, MW, concentration, and irradiation doses are required to make biocompatible CMHA hydrogels [3,4]. In this particular study, the DS and MW of the CMHA produced were 0.87 and 149 kDa, respectively. This will be used to investigate milder irradiation conditions for the formation of hydrogels by combining CMHA with linear and slightly crosslinked PAA. The sections that follow discuss the characterization of the polymer blends, such as their gel properties, molecular structure, thermal stability, and in-vitro degradation using PBS (pH 7.4).

3.2. Gel Properties of CMHA-PAA and CMHA-Carbopol Blends

Hydrogels for biomedical uses undergo initial characterization through the determination of their gel content and swelling properties. These two properties can be used as a means to assess the stability of hydrogels in aqueous solutions [9]. The gel properties of CMHA-PAA are shown in Figure 1. The gel fractions substantially increased as the absorbed dose became higher, from 20 to 40 kGy (Figure 1a). Hydrogel blends of 35% CMHA, 5% PAA had higher gel fractions compared to the 37.5% CMHA, 2.5% PAA blend for both doses. These values show that higher irradiation doses and higher concentrations of PAA result in more crosslinking, hence the higher gel fractions.

Different trends for the degree of swelling were observed as the dose increased (Figure 1b). The degree of swelling of the 35% CMHA, 5% PAA blend decreased from 1115 to 616 g/g, while the formulation with 37.5% CMHA, 2.5% PAA increased from 522 to 895 g/g. From the results of the gel fraction experiment, higher degree of swelling are expected for the hydrogels with more PAA and those that are irradiated at higher doses, due to more crosslinking. However, the swelling ratio obtained for the 35% CMHA, 5% PAA did not follow the same trend at 40 kGy, possibly because the higher degree of crosslinks caused the hydrogel to be more rigid, hence restricting its ability to swell and hold water in its network.

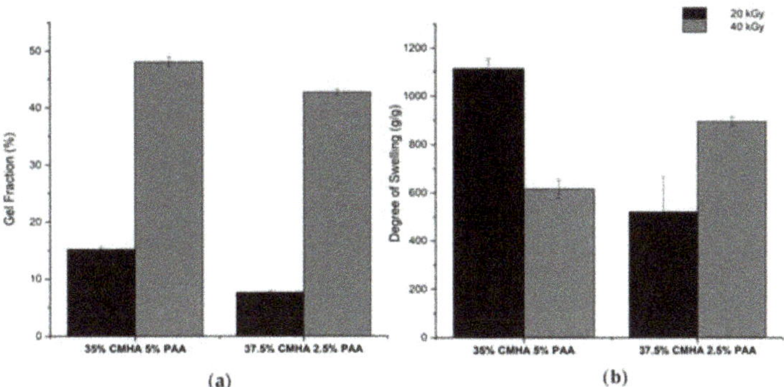

Figure 1. Gel fraction (**a**) and degree of swelling (**b**) of CMHA-PAA hydrogels.

For the CMHA-Carbopol hydrogel blends, it was also observed that gel fractions increased proportionally with the dose for all samples (Figure 2a). However, the gel fractions decreased when CMHA was mixed with Carbopol, with the 10% pure Carbopol samples obtaining gel fractions in the range of 95–99%, while the 10% Carbopol, 10% CMHA obtained 60–81%. The soluble fraction from the 10% Carbopol, 10% CMHA hydrogel may have predominantly originated from the degraded CMHA chains. This could be due to the fact that some of the CMHA polymeric chainswere degraded since polysaccharides, including hyaluronic acid and their derivatives, are susceptible to chain scission when exposed to high-energy radiation. This radiation-induced chain scission is more apparent at low CMHA concentrations (i.e., 10% CMHA).

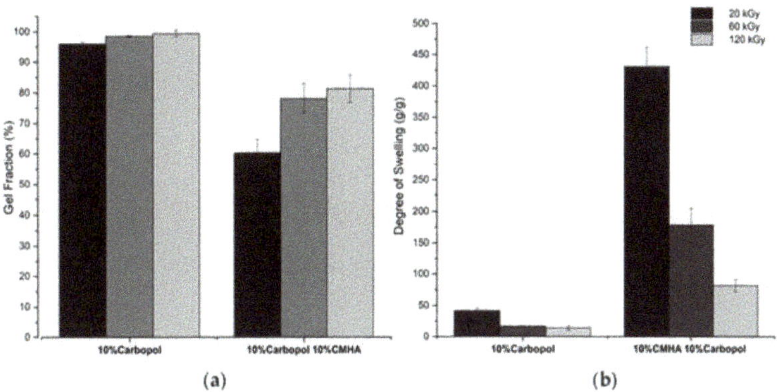

Figure 2. Gel fraction (**a**) and degree of swelling (**b**) of CMHA: Carbopol blends.

As previously discussed, the degree of swelling decreases for samples with a high PAA concentration at higher doses due to the formation of more crosslinks within the hydrogel (Figure 2b). Higher degrees of swelling of 10% Carbopol, 10% CMHA hydrogel blends were observed compared to the pure 10% Carbopol samples, which may have been caused by the presence of more flexible polymer chains having a network of lesser crosslinks.

The hydrogel blends based on CMHA and polyacrylic acid polymers produce interesting gel properties that can be controlled by modifying either the polymer concentrations, the irradiation dose, or both. It is also possible to produce gels using milder irradiation conditions by considering concentrations with an equal or lower weight ratio of CMHA to PAA polymers (1:1 or lower than 7:1).

3.3. FT-IR and TGA Analysis

One way to determine the successful crosslinking or blending of polymer chains within the hydrogel is by analyzing their FT-IR spectra and comparing them from their pristine raw materials [10]. The FT-IR characteristic peaks of the raw materials HA, CMHA, PAA, and Carbopol, as well as the CMHA-PAA hydrogel blends, are shown in Figure 3. The 35% CMHA, 5% PAA hydrogel irradiated at 40 kGy has shown more resemblance to the spectrum of its dominant polymer CMHA, but with a new peak around 1705 cm^{-1} due to the C=O carboxylic group of PAA. The FT-IR spectra of 10% CMHA, 10% Carbopol irradiated at 20 kGy showed a significant deviation from its pristine material. Specifically, the C=O carboxylate group from PAA showed a strong peak at 1410 cm^{-1}, obscuring the C=O carboxylic groups of hyaluronic acid as well as the decrease in peak intensity at 1040 cm^{-1} corresponding to the C-OH alcohol moiety from CMHA. It was also observed that the change in the shape of the characteristic bands of -OH at 3000–2850 cm^{-1} changes depending on the weight ratio of the polymers.

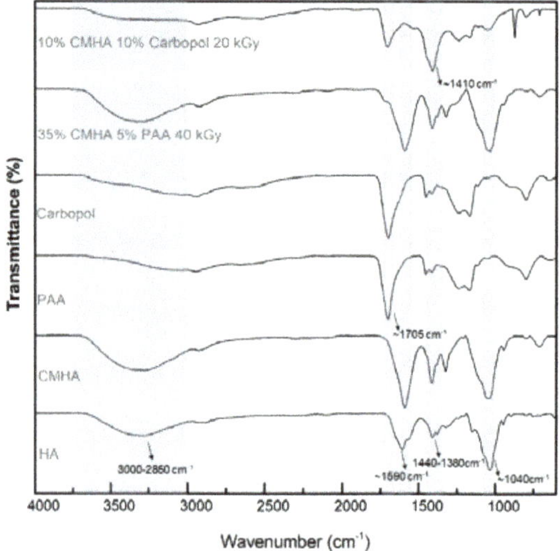

Figure 3. FT-IR spectra HA, CMHA, crosslinked CMHA, crosslinked CMHA-PAA, and crosslinked CMHA-Carbopol hydrogels.

The thermograms of the CMHA-PAA and CMHA-Carbopol hydrogels without soluble fractions are shown in Figure 4. The TG and DTG profiles depicted the multistep decomposition of the samples. The possible dehydration of 35% CMHA, 5% PAA 40 kGy was 71.7 °C, which is generally lower than the 164.5 °C of 10% CMHA, 10% Carbopol 20 kGy hydrogels (Figure 4). Previous studies have also observed a great difference in the dehydration temperature between their PAA-based samples [11]. At 185–315 °C, the decarboxylation and depolymerization of both the PAA and CMHA main chains of the hydrogel blends occur [4,12]. The observed mass losses greater than 315 °C were potentially from crosslinked CMHA chains, CMHA-PAA chains, andPAA chains with different crosslink densities formed after irradiation [13].

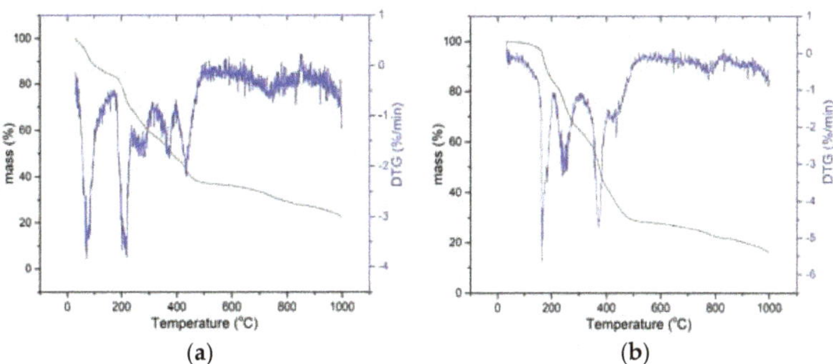

Figure 4. TGA and DTG profiles of (**a**) 35% CMHA, 5% PAA irradiated at 40 kGy and (**b**) 10% CMHA, 10% Carbopol irradiated at 20 kGy hydrogels.

3.4. In-Vitro Biodegradation of Hydrogels in Phosphate Buffer Solution (PBS)

By using the simulated ionic and pH conditions of biological fluids, such as PBS (pH = 7.4), at 37 °C, the stability of the samples could be initially assessed [14]. Figure 5 shows that a formulation with the equal weight ratio of 10% CMHA, 10% Carbopol irradiated in 20 kGy has a stable crosslinking network exhibiting a stable swelling property without a significant increase in the degradation percentage of PBS. Biomedical hydrogels with stable gel properties in biological fluids can potentially be used for applications that need longer exposure when implanted in the body.

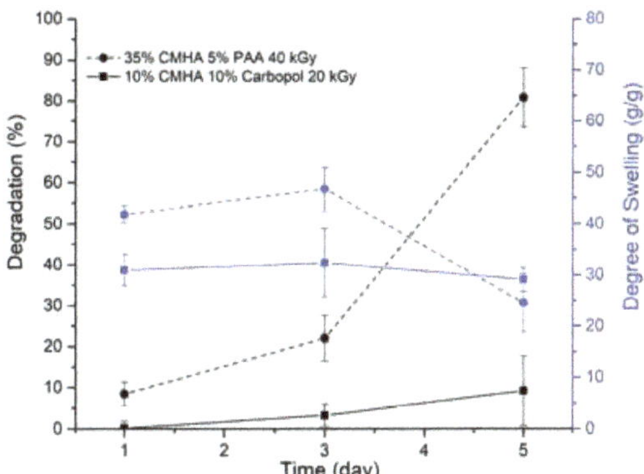

Figure 5. In-vitro degradation and swelling properties of CMHA with poly(acrylic) acid hydrogels in PBS (pH 7.4) at 37 °C.

4. Conclusions

Semi-synthetic hydrogels based on CMHA with polyacrylic acid (CMHA-PAA) and CMHA-Carbopol) were produced using electron beam irradiation without the use of toxic initiators or crosslinkers. The gel properties of the produced gels can easily be modified by changing the polymer weight ratio and the absorbed radiation dose according to the results. Crosslinking between CMHA and PAA polymer chains was assessed by changes in the IR spectra and thermal stability of the samples. Moreover, the in-vitro biodegradability study showed the stability of the produced hydrogels. Since hyaluronic acid is ubiquitous to the human body, CMHA-based hydrogels have a strong potential for use in biomedical applications.

Author Contributions: All authors contributed to the study conception and design. Material preparation, data collection, and analysis were performed by A.K.R.G. and A.P.S. Study investigation and supervision were led by L.S.R. The first draft of the manuscript was written by A.K.R.G. and all authors commented on previous versions of the manuscript. All authors have read and approved the final manuscript. All authors have read and agreed to the published version of the manuscript.

Funding: This research received no external funding.

Institutional Review Board Statement: Not applicable.

Informed Consent Statement: Not applicable.

Data Availability Statement: The data presented in this study are available upon request from the corresponding author.

Conflicts of Interest: The authors declare no potential conflicts of interest with respect to the research, authorship, and/or publication of this article.

References

1. Mignon, A.; De Belie, N.; Dubruel, P.; Van Vlierberghe, S. Superabsorbent polymers: A review on the characteristics and applications of synthetic, polysaccharide-based, semi-synthetic and 'smart' derivatives. *Eur. Polym. J.* **2019**, *117*, 165–178. [CrossRef]
2. Rosiak, J.M.; Yoshii, F. Hydrogels and their medical applications. *Nucl. Instrum. Methods Phys. Res. Sect. B Beam Interact. Mater. Atoms* **1999**, *151*, 56–64. [CrossRef]
3. Relleve, L.S.; Gallardo, A.K.R.; Abad, L.V. Radiation crosslinking of carboxymethyl hyaluronic acid. *Radiat. Phys. Chem.* **2018**, *151*, 211–216. [CrossRef]
4. Relleve, L.S.; Gallardo, A.K.R.; Tecson, M.G.; Luna, J.A.A. Biocompatible hydrogels of carboxymethyl hyaluronic acid prepared by radiation-induced crosslinking. *Radiat. Phys. Chem.* **2021**, *179*, 109194. [CrossRef]
5. Gallardo, A.K.R.; Relleve, L.S.; Barba, B.J.D.; Cabalar, P.J.E.; Luna, J.A.A.; Tranquilan-Aranilla, C.; Madrid, J.F.; Abad, L.V. Application of factorial experimental design to optimize radiation-synthesized and biodegradable super water absorbent based on cassava starch and acrylic acid. *J. Appl. Polym. Sci.* **2021**, *139*, 51451. [CrossRef]
6. Shin, M.S.; Kim, S.J.; Park, S.J.; Lee, Y.H.; Kim, S.I. Synthesis and characteristics of the interpenetrating polymer network hydrogel composed of chitosan and polyallylamine. *J. Appl. Polym. Sci.* **2002**, *86*, 498–503. [CrossRef]
7. Nho, Y.C.; Park, J.S.; Lim, Y.M. Preparation of poly(acrylic acid) hydrogel by radiation crosslinking and its application for mucoadhesives. *Polymers* **2014**, *6*, 890–898. [CrossRef]
8. Biswas, G.R.; Majee, S.B.; Roy, A. Combination of synthetic and natural polymers in hydrogel: An impact on drug permeation. *J. Appl. Pharm. Sci.* **2016**, *6*, 158–164. [CrossRef]
9. Edsman, K.; Nord, L.I.; Öhrlund, Å.; Lärkner, H.; Kenne, A.H. Gel properties of hyaluronic acid dermal fillers. *Dermatol. Surg.* **2012**, *38*, 1170–1179. [CrossRef] [PubMed]
10. Grabowska, B.; Kaczmarska, K.; Bobrowski, A.; Kurleto-Kozioł, Z.; Szymański, L. Crosslink the Novel Group of Polymeric Binders BioCo by the UV-radiation. *Arch. Foundry Eng.* **2016**, *16*, 85–88. [CrossRef]
11. Dil, N.N.; Sadeghi, M. Free radical synthesis of nanosilver/gelatin-poly (acrylic acid) nanocomposite hydrogels employed for antibacterial activity and removal of Cu(II) metal ions. *J. Hazard. Mater.* **2018**, *351*, 38–53. [CrossRef] [PubMed]
12. Fan, X.D.; Hsieh, Y.L.; Krochta, J.M.; Kurth, M.J. Study on molecular interaction behavior, and thermal and mechanical properties of polyacrylic acid and lactose blends. *J. Appl. Polym. Sci.* **2001**, *82*, 1921–1927. [CrossRef]
13. Relleve, L.S.; Aranilla, C.T.; Barba, B.J.D.; Gallardo, A.K.R.; Cruz, V.R.C.; Ledesma, C.R.M.; Nagasawa, N.; Abad, L.V. Radiation-synthesized polysaccharides/polyacrylate super water absorbents and their biodegradabilities. *Radiat. Phys. Chem.* **2020**, *170*, 108618. [CrossRef]
14. Deshmukh, M.; Singh, Y.; Gunaseelan, S.; Gao, D.; Stein, S.; Sinko, P.J. Biodegradable poly (ethylene glycol) hydrogels based on a self-elimination degradation mechanism. *Biomaterials* **2010**, *31*, 6675–6684. [CrossRef] [PubMed]

Proceeding Paper

Review on Thermal Energy Storing Phase Change Material-Polymer Composites in Packaging Applications [†]

Tejashree Amberkar * and Prakash Mahanwar

Department of Polymer and Surface Engineering, Institute of Chemical Technology, Mumbai 400019, India;
pa.mahanawar@ictmumbai.edu.in
* Correspondence: tejuamberkar@yahoo.co.in
† Presented at the 2nd International Online Conference on Polymer Science—Polymers and Nanotechnology for
Industry 4.0, 1–15 November 2021; Available online: https://iocps2021.sciforum.net/.

Abstract: Thermally sensitive food and pharma packages are maintained at desired temperatures using refrigeration systems. These systems are powered by non-conventional energy resources. They provide uneven cooling in large containers. Interruption in their functioning during supply chain activities increases their energy requirements. Studies revealed that using phase change material (PCM)-polymer composites in refrigeration systems and packaging containers curtailed energy utilization for maintaining a consistent temperature. These composites maintain a temperature around its phase change temperature by absorbing or releasing latent heat. This review discusses different designs of PCM-polymer composites that maintain the temperature of big shipments and small containers.

Keywords: phase change material; thermal energy storage; latent heat

1. Introduction

Temperature-controlled packaging is a high-growth sector with a predicted compounded annual growth rate of 18.14% until 2026 [1]. The surge in demand is mainly expected to result from temperature-controlled vaccine packages, biologics, and small e-commerce packages. During the COVID-19 pandemic, the increased demand for temperature-controlled packaging is for biopharma products. The traditional methods for controlling shipment temperatures involve active and passive temperature control methods. Active systems consist of cooling arrangements facilitated by electricity or fuel. The excessive use of these systems results in the consumption of high amounts of non-renewable energy, which ultimately impacts the environment in the long term. One more disadvantage of active systems is temperature excursion during transport activities. These activities, such as transferring goods to carriers at the shipping dock or airport, improper handling of goods by unskilled labor, schedule changes, excessive or low amounts of coolant, mechanical damage, etc., can vary temperatures beyond the decided limits. Such disrupted supply chain systems can spoil food and lifesaving pharma products, such as vaccines and biologics.

On the other hand, passive systems are more energy-efficient and environmentally friendly for storing temperature-sensitive products than active systems. Passive systems use coolants, such as ice packs, gel packs, and PCMs, along with insulation material to maintain the required temperature. PCMs can absorb, store, or release latent heat while undergoing phase transition and maintain the product in the predetermined temperature range. PCMs incorporated in refrigerated active packaging systems have maintained temperatures at the desired levels for as long as ten days in the absence of electricity [2]. PCM-incorporated refrigeration systems have also shown reduced temperature fluctuations. This reduced power demand results in energy savings. PCMs in conjunction with insulators can be used in mobile vaccine, food, and e-commerce containers. These passive containers can be charged once in the phase transition temperature range and can be used

Citation: Amberkar, T.; Mahanwar, P. Review on Thermal Energy Storing Phase Change Material-Polymer Composites in Packaging Applications. *Mater. Proc.* **2021**, *7*, 14. https://doi.org/10.3390/IOCPS2021-11218

Academic Editor: Shin-ichi Yusa

Published: 25 October 2021

Publisher's Note: MDPI stays neutral with regard to jurisdictional claims in published maps and institutional affiliations.

for a couple of hours without electricity. PCM-incorporated shippers decrease the cost of smaller shipments and achieve optimum performance. To understand the workings of PCM-incorporated packaging systems in detail, this paper is divided into two sections. The first section is dedicated to innovations using PCMs in increasing the refrigeration efficiency of large container shipments. The second section provides information about recent developments in small container packages.

2. Large Container Shipments

Refrigerator vehicle trucks are commonly used for transporting thermally sensitive goods in every part of the globe. However, variations in temperature across different territories and times of the day increase the power requirements of refrigeration. Higher temperature variation between the external and internal walls of the container increases the number of compression cycles and reduces its operation time. Such a working style necessitates frequent replacement of the compressor. Refrigerants used in compressors are greenhouse gases; thus, the increased use of compressors poses a significant risk of greenhouse gas leakage in the environment. Further, a significant amount of energy is spent on operating refrigerators with high temperature gradients between external and internal environments. Increasing the efficiency of refrigerator systems will be helpful for the environment. PCMs used in the walls of large shipment containers, such as refrigerated trucks and bulk pallet shippers, have increased energy efficiency by significant levels. The placement of PCM cold plates for trucks is as shown in Figure 1. Many researchers observed improvements in the thermoregulation of packaged goods by incorporating PCM plates in big shippers. Thus, PCM-incorporated bulk shippers are commercialized and used for transporting thermosensitive products. Though the initial investment cost for these shippers is high, the assembly has proven to be cost-effective for long-term usage over conventional shippers. PCM RT 5 was inserted into cold storage plates [3]. Each cold plate carried 126 kg of PCMs. Nineteen of such cold storage plates were placed on the roof, and one was placed in the upper part of the front wall side of the refrigerated container, with dimensions as per ISO 40. The container was insulated with 100 mm of polyurethane foam. The refrigerant passed through the fin tube to charge the PCMs. It took 6 h to fully charge a PCM-stored cold storage plate to its phase transition temperature. After that, it maintained the temperature below 12 °C for 14 long hours without using a diesel-run refrigeration system. This system was more expensive than diesel-run refrigeration systems due to the high cost of PCMs. However, its operation cost was 61.9% less than conventional systems. This means that the initial high-cost payback is 0.58 years. The container was flexible to use on the road and rail tracks.

Figure 1. Placement of PCM plates in a truck.

A truck with two chambers filled with different PCMs was constructed [4]. The compartments had PCMs with phase transition temperatures of 2.34 °C and −15 °C. The compartments maintained the phase change temperature without an external power supply for 9.2 h and 6.2 h, respectively. A six-ton truck was equipped with six thin PCM cold plates [5]. Each plate consisted of 35 kg of PCMs. The cooling performance of three different PCMs was tested. The E26 PCM performed superiorly. It provided a high melting time of 17,200 s at a truck speed of 81 km per hour and 18,400 s for a stationary truck. The moving truck increased heat transfer and reduced melting time. The PCM maintained the temperature for a 491 km distance at a 110 km per hour truck speed. Thus, it helped in minimizing the use of refrigeration systems. Radebe and Huan reported that PCM eutectic plates with a salt-water solution could be incorporated into trucks transporting agricultural goods for temperature maintenance [6]. The use of PCMs maintained the temperature inside of the truck to the desired level, thus preventing the degradation of agro products.

Principi et al. [7] used PCMs in two ways to lower energy consumption. The team had incorporated PCMs with a phase transition temperature of 35 °C near the outer boundary of the refrigerated truck. The PCM layer acted as a thermal buffer and prevented solar heat from reaching the inner surface of the truck. The maximum time delay for the heat to reach the interior was 4.3 h. It allowed 8.57% less heat to reach the truck's interior than the control reference. Heat reduction curtailed the refrigerator's energy requirement for maintaining the interior temperature; this is the reason for the efficient energy consumption of the PCM-incorporated heat exchanger refrigerator. During the OFF time of the compressor, the heat was absorbed by the PCMs to liquefy at 5 °C. During the ON time, the heat released by the PCMs was absorbed by evaporator outlet air. Therefore, during the OFF time of the compressor, the PCMs maintained the temperature longer. The PCM freezing process increased the start time of the compressor. The summation of these two effects resulted in a lower number of compressor cycles and increased the duration of the operation. This change in the cycles' working reduced the energy demand of the compressor by 16%. Fioretti et al. [8] studied the effect of adding PCM panels near the outer boundary of the cold room. The thermal testing carried out on the prototype helped to determine the PCM's performance in actual reefer containers. In this test, PCM panels along with polyurethane foam sheets were enclosed within metal sheets, and their thermal performance results were compared with the reference control sample. The reference panel did not contain PCM sheets. The arrangement of the prototype panel can be better understood using Figure 2. When these panels were attached to cold room walls on external sides, the resultant heat reduction saved energy by 4.7%.

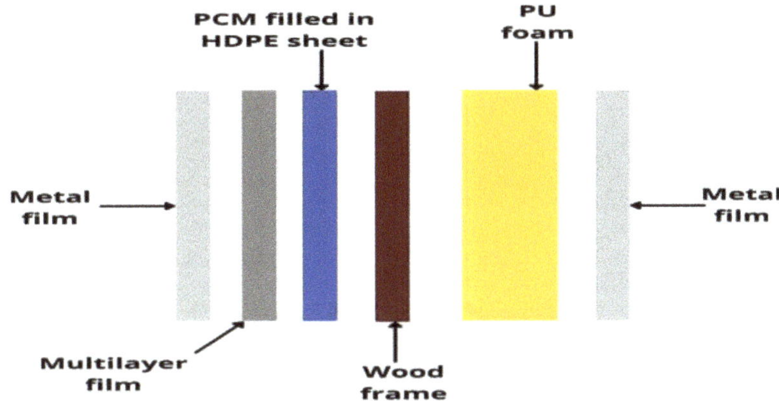

Figure 2. PCM panel.

The transport air conditioning systems' efficiency levels were improved by using serrated fins on the air side and perforated straight fins near the PCM [9]. The discharging performance is the indicator of the temperature maintenance of the system. The designed device had a discharging depth of more than 97% and could cool down the environment in seconds. The compactness and high heat transfer performance of the system will be beneficial for use in reefers. A compact PCM-incorporated air conditioner (AC) design was created for space-sensitive transportation refrigeration systems [10]. The assembly consisted of rectangular straight perforated fins in PCM chambers. Air channels with serrated fins were positioned orthogonally to the PCM chamber. Both the chamber and channel were periodically connected to the clapboard to provide a compact structure. The structure provided emergency cooling at a rate nine times higher than conventional AC systems and reduced temperature fluctuation to a lower value of 2.56 °C compared to 4.3 °C for conventional systems.

3. Small Container Packages

PCMs or microencapsulated phase change materials (MPCMs) are filled in rigid containers or flexible pouches. These PCM slabs can be used with or without insulation material in small containers to maintain the temperature of packaged products without electricity. These assemblies can be better understood using Figure 3.

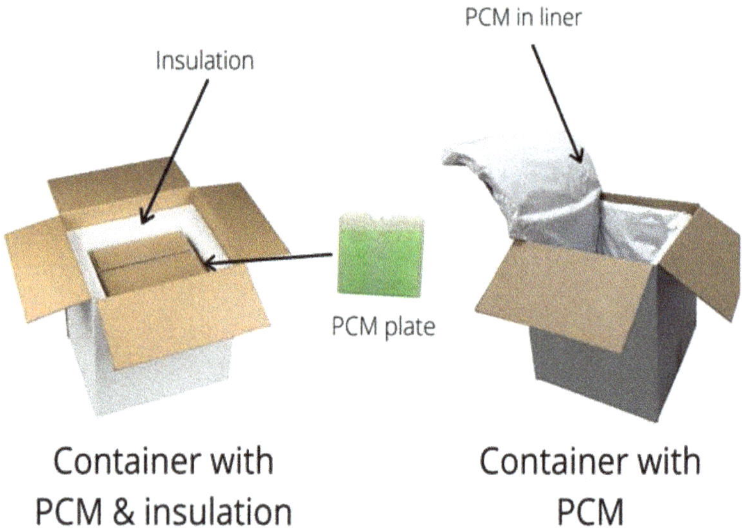

Figure 3. Small container packages with PCM plates and insulation versus PCM liners only.

These assemblies maintain the temperature of packaged products near the phase transition temperature of the PCM. The literature, in which PCM-incorporated small containers' heat transfer performance is studied, is discussed in this section. In one study, RT 6 PCM was encapsulated into porous calcium silicate [11]. This shape-stabilized PCM had a high melting enthalpy of 174 J/g at 8 °C. It maintained the shipment temperature at the desired level for 4–9 h in the ambient atmosphere. Sodium chloride hydrate, along with a nucleating agent and other additives, was encapsulated in plastic brick [12]. When ice cream stored at −24 °C was kept in an ambient atmosphere of 20 °C, packaging with PCM + insulation, only insulation, and a control sample showed a temperature rise of less than 1 °C, 9 °C, and 42 °C, respectively, in 40 min near the surface. This confirmed the superiority of PCMs over insulation material in temperature-controlled packaging systems. In another study, octanoic acid was microencapsulated in the polystyrene shell and incorporated in chocolate shipper walls [13]. It maintained the desired temperature for 6–8.8 h. Yie et al. [14] prepared a silica aerogel-PCM composite structure. The porous structure of the aerogel was filled with microencapsulated PCMs using the impregnation method. This composite, when combined with the insulator board, increased the temperature maintenance period by 99 times. Xu et al. [15] prepared a container for storing apples in the temperature range of 2–8 °C with PCMs and insulation. The addition of PCMs maintained the temperature for 9.63 h. Without PCMs, this time was 0.77 h. Wang et al. [16] studied thermal buffering characteristics of meat packaging with polystyrene-shelled PCM microcapsules. It maintained the meat temperature at the desired level for 30 min.

Huang and Pinolek [17] designed a container that combined polyurethane insulation, vacuum insulation panels, and thermal energy-storing PCM panels to maintain the desired temperature for more than 72 h within the range of 2–8 °C in varying ambient temperature ranges from −20 °C to 35 °C. Buska [18] designed a cup whose walls were filled with PCMs to maintain the beverage temperature at the desired level. It helped in consuming the beverage at the required temperature level for a longer time with less energy used.

4. Conclusions

Due to limited fossil resources and the increased need for environmentally friendly, sustainable technologies, the importance of using PCMs to reduce thermal energy waste will increase in decades to come. The culture of using PCMs in packaging is growing exponentially. PCM consumption seems to be an emerging trend in various fields, such as e-commerce packaging, food packaging, and pharma packaging. This paper provides information about different packaging systems utilizing PCMs for transporting temperature-sensitive products. The contribution of traditional cooling systems, such as AC and ice water systems, is contracting mainly due to their higher cost for smaller shipments. A steadily growing knowledge base has demonstrated that PCMs can replace traditional cooling systems and even improve their performance.

Supplementary Materials: The following are available online at https://www.mdpi.com/article/10.3390/IOCPS2021-11218/s1.

Author Contributions: Writing—original draft preparation, T.A.; writing—review and editing, P.M. All authors have read and agreed to the published version of the manuscript.

Funding: This work is funded by the AICTE National Doctoral Fellowship Scheme sanctioned vide letter F No.: 12-2/2019-U1 provided by the Ministry of Human Resource Development, Government of India.

Institutional Review Board Statement: Not applicable.

Informed Consent Statement: Not applicable.

Data Availability Statement: Since this is review article data is taken from literature.

Acknowledgments: The authors acknowledge library facilities provided by the Institute of Chemical Technology, Mumbai, in accessing research databases.

Conflicts of Interest: The authors declare no conflict of interest.

References

1. Temperature Controlled Packaging Solutions Market by Type (Active, Passive), Product, Usability (Single, Reuse), Revenue type (Product, Service), End-Use Industry (Pharma and Biopharma) & Region—Trends and Forecasts Up to 2026. Available online: https://www.marketsandmarkets.com/Market-Reports/temperature-controlled-packaging-solutions-market-522770 1.html (accessed on 29 September 2021).
2. Vaccine Fridge Keeps Its Cool during 10-Day Power Cut. Available online: https://www.newscientist.com/article/mg20927944-000-vaccine-fridge-keeps-its-cool-during-10-day-power-cut (accessed on 29 September 2021).
3. Tong, S.; Nie, B.; Li, Z.; Jin, Y.; Ding, Y.; Hu, H. Investigation of the cold thermal energy storage reefer container for cold chain application. *Energy Storage Sci. Technol.* **2020**, *9*, 211–216. [CrossRef]
4. Zhang, S.; Zhang, X.; Xu, X.; Zhao, Y. Thermomechanical Analysis and Numerical Simulation of Storage Type Multi-Temperature Refrigerated Truck. *Int. J. Sci.* **2020**, *7*, 261–270.
5. Mousazade, A.; Rafee, R.; Valipour, M.S. Thermal performance of cold panels with phase change materials in a refrigerated truck. *Int. J. Refrig.* **2020**, *120*, 119–126. [CrossRef]
6. Radebe, T.B.; Huan, Z.; Baloyi, J. Simulation of eutectic plates in medium refrigerated transport. *J. Eng. Des. Technol.* **2021**, *19*, 62–80. [CrossRef]
7. Principi, P.; Fioretti, R.; Copertaro, B. Energy saving opportunities in the refrigerated transport sector through Phase Change Materials (PCMs) application. *J. Phys. Conf. Ser.* **2017**, *923*, 12043. [CrossRef]
8. Fioretti, R.; Principi, P.; Copertaro, B. A refrigerated container envelope with a PCM (Phase Change Material) layer: Experimental and theoretical investigation in a representative town in Central Italy. *Energy Convers. Manag.* **2016**, *122*, 131–141. [CrossRef]

9. Nie, B.; She, X.; Zou, B.; Li, Y.; Li, Y.; Ding, Y. Discharging performance enhancement of a phase change material based thermal energy storage device for transport air-conditioning applications. *Appl. Therm. Eng.* **2020**, *165*, 114582. [CrossRef]

10. Nie, B.; She, X.; Du, Z.; Xie, C.; Li, Y.; He, Z.; Ding, Y. System performance and economic assessment of a thermal energy storage based air-conditioning unit for transport applications. *Appl. Energy* **2019**, *251*, 113254. [CrossRef]

11. Johnston, J.H.; Grindrod, J.E.; Dodds, M.; Schimitschek, K. Composite nano-structured calcium silicate phase change materials for thermal buffering in food packaging. *Curr. Appl. Phys.* **2008**, *8*, 508–511. [CrossRef]

12. Leducq, D.; Ndoye, F.T.; Charriau, C.; Alvarez, G. Thermal protection of ice cream during storage and transportation. *Refrig. Sci. Technol.* **2015**, 4614–4619. [CrossRef]

13. Ünal, M.; Konuklu, Y.; Paksoy, H. Thermal buffering effect of a packaging design with microencapsulated phase change material. *Int. J. Energy Res.* **2019**, *43*, 4495–4505. [CrossRef]

14. Yin, H.; Gao, S.; Cai, Z.; Wang, H.; Dai, L.; Xu, Y.; Liu, J.; Li, H. Experimental and numerical study on thermal protection by silica aerogel based phase change composite. *Energy Rep.* **2020**, *6*, 1788–1797. [CrossRef]

15. Xu, X.; Zhang, X.; Zhou, S.; Wang, Y.; Lu, L. Experimental and application study of Na $_2$ SO $_4$ ·10H $_2$ O with additives for cold storage. *J. Therm. Anal. Calorim.* **2019**, *136*, 505–512. [CrossRef]

16. Wang, Y.; Zhang, Q.; Bian, W.; Ye, L.; Yang, X.; Song, X. Preservation of traditional Chinese pork balls supplemented with essential oil microemulsion in a phase-change material package. *J. Sci. Food Agric.* **2020**, *100*, 2288–2295. [CrossRef] [PubMed]

17. Huang, L.; Piontek, U. Improving performance of cold-chain insulated container with phase change material: An experimental investigation. *Appl. Sci.* **2017**, *7*, 1288. [CrossRef]

18. Booska, R. Thermal Receptacle with Phase Change Material. U.S. Patent 10,595,654, 24 March 2020.

materials proceedings

Abstract

A Novel 3D Microporous Structure Hydrogel with Stable Mechanical Properties and High Elasticity and Its Application in Sensing [†]

Chunyin Lu, Jianhui Qiu *, Eiichi Sakai and Guohong Zhang

Department of Machine Intelligence and Systems Engineering, Faculty of Systems Science and Technology, Akita Prefectural University, Akita 015-0055, Japan; d21s005@akita-pu.ac.jp (C.L.); e_sakai@akita-pu.ac.jp (E.S.); zgh131523@163.com (G.Z.)
* Correspondence: Jianhui_Qiu@akita-pu.ac.jp
† Presented at the 2nd International Online Conference on Polymer Science—Polymers and Nanotechnology for Industry 4.0, 1–15 November 2021; Available online: https://iocps2021.sciforum.net/.

Abstract: Hydrogels have recently been increasingly studied due to their similarity to natural soft tissues. However, the stable mechanical properties and elasticity required for hydrogels used in sensing and wearable devices remain challenging. Herein, a novel 3D microporous structure hydrogel with favorable stable mechanical properties and elasticity is developed via a simple and economical method. The good resilience (94.5%) and lower residual strain (11.5%) are realized based on the results of 20 successive cycles at a strain of 300%. The elasticity of the hydrogel is achieved by varying the effective network chain density. The prepared hydrogel has stable mechanical properties and a high elasticity, resulting in remarkable performance when used in sensors. The hydrogel-based sensors can accurately and consistently record human activities when used as wearable sensors. This work provides a new way to simply and effectively prepare hydrogels, which has great potential to be widely applicated in sensing and flexible devices, such as health-recording sensors, wearable devices, and artificial intelligence.

Keywords: 3D microporous structure; hydrogel; sensor; stable mechanical properties; elasticity

Citation: Lu, C.; Qiu, J.; Sakai, E.; Zhang, G. A Novel 3D Microporous Structure Hydrogel with Stable Mechanical Properties and High Elasticity and Its Application in Sensing. *Mater. Proc.* **2021**, *7*, 15. https://doi.org/10.3390/IOCPS2021-11212

Academic Editor: Melkie Tadesse

Published: 20 October 2021

Publisher's Note: MDPI stays neutral with regard to jurisdictional claims in published maps and institutional affiliations.

Supplementary Materials: The supporting information can be downloaded at: https://www.mdpi.com/article/10.3390/IOCPS2021-11212/s1.

Institutional Review Board Statement: Not applicable.

Informed Consent Statement: Not applicable.

Data Availability Statement: Not applicable.

materials proceedings

MDPI

Proceeding Paper

Assessment of Recycled PLA-Based Filament for 3D Printing [†]

Antonella Patti [1,*], Stefano Acierno [2], Gianluca Cicala [1], Mauro Zarrelli [3] and Domenico Acierno [4,*]

[1] Department of Civil Engineering and Architecture (DICAr), University of Catania, Viale Andrea Doria 6, 95125 Catania, Italy; gianluca.cicala@unict.it

[2] Department of Engineering, University of Sannio, Piazza Roma 21, 82100 Benevento, Italy; stefano.acierno@unisannio.it

[3] Institute of Polymers, Composites and Biomaterials, National Research Council, P. le Enrico Fermi 1, 80055 Naples, Italy; mauro.zarrelli@cnr.it

[4] CRdC Nuove Tecnologie per le Attività Produttive Scarl, Via Nuova Agnano 11, 80125 Naples, Italy

[*] Correspondence: antonella.patti@unict.it (A.P.); acierno@crdctecnologie.it (D.A.)

[†] Presented at the 2nd International Online Conference on Polymer Science—Polymers and Nanotechnology for Industry 4.0, 1–15 November 2021; Available online: https://iocps2021.sciforum.net/.

Abstract: This study investigated the possibility of replacing virgin matrices with recycled polymers in additive manufacturing (AM). In this regard, two commercial filaments, made from polylactide acid (PLA)—the second (here referred to as recycled) obtained from the recovery of waste production of the first one (here referred to as virgin)—were initially characterized using infrared (IR) spectroscopy, thermogravimetric analysis (TGA) and dynamic rheology. Then, filaments were extruded in a 3D printer and characterized by dynamic mechanical analysis (DMA). Despite a small reduction in the intensity of correspondence of typical absorption bands of the PLA polymer, in the case of the recycled material compared to the virgin one (as attested by IR spectra), the thermal-mechanical results allow us to attest the very similar characteristics of recycled and neat filaments. The onset of thermal degradation was found at around 315 °C in both systems. Both materials exhibited the same frequency- and time-dependent trends of the complex viscosity, with a reduction of approximately 30% after 10 min of testing. When samples were dried at 80 °C under vacuum for 10 h, the stabilization of rheological features against time was improved. There was no significant difference in the storage modulus (E′) and dissipation factor (tan delta) of 3D printed parts made with different types of PLA-based filaments.

Keywords: poly(lactide) acid; 3D printing; recycling; thermo-mechanical properties; rheological characterization

Citation: Patti, A.; Acierno, S.; Cicala, G.; Zarrelli, M.; Acierno, D. Assessment of Recycled PLA-Based Filament for 3D Printing. *Mater. Proc.* **2021**, *7*, 16. https://doi.org/10.3390/IOCPS2021-11209

Academic Editor: Francesca Lionetto

Published: 20 October 2021

Publisher's Note: MDPI stays neutral with regard to jurisdictional claims in published maps and institutional affiliations.

1. Introduction

Plastics are extremely useful for a wide range of applications due to their mechanical and chemical properties, as well as their ease of manipulation [1]. Yet, not being biodegradable, plastic materials pose a serious environmental problem due to the accumulation of products in nature [2]. This aspect has become particularly relevant in the sustainable development of industrial production [3].

Nonetheless, additive manufacturing (AM), well known as 3D printing, is emerging as a crucial industrial technology for rapid prototyping, in order to convert a numerical model into material deposition and 3D printed parts [4]. During this cycle, a huge amount of waste products is developed. In order to reduce plastic waste [5] and limit the environmental impact of the AM process [6], bio-based and recycled polymers have been considered as alternatives to conventional raw materials.

Polylactic acid (PLA), an aliphatic polyester derived from 100% renewable resources, represents a common thermoplastic polymer most often utilized in the AM field, taking into account its excellent biocompatibility and environmental sustainability, absence of unpleasant odors during handling and realization of final products with fair precision tolerance [5].

Although PLA is considered one of the most promising bio-based alternatives to common non-biodegradable polymers from petroleum [6], its massive use could generate an important waste accumulation [7]. End-of-life disposal of such materials should be carefully considered in order to facilitate the transition to a circular economy for plastics [8]. Mechanical recycling of PLA is a process consisting of the collection, washing and reprocessing of thermoplastic materials, usually achieved thermally, with the main purpose of preserving the original properties [9]. This recycling method has received a significant amount of attention because it is relatively simple, requires little investment and its technological parameters are controlled [10]. During recycling, the major problem is the thermal stability of PLA and the slight decrease in mechanical properties after several extrusion processes [7].

In this framework, this study was focused on improving the sustainability aspects of the AM technology by verifying the thermal and mechanical characteristics of recycled polymers, coming from waste products, in comparison with virgin matrices, for developing 3D printed parts.

2. Materials and Methods

This experiment used commercially available filaments made from a poly(lactide) acid (PLA)-based polymer. In particular, a basic PLA, here referred to as virgin PLA (cod. Ingeo™ 4043D, density of 1.24 g/cc ASTM D792, MFR of 6 g/10 min ASTM D1238, a product from NatureWorks LLC, Minnetonka, MN, USA), and a recycled filament derived from the production waste of the first filament, here referred to as recycled PLA, were supplied by EUMAKERS (Barletta, Italy).

On these filaments, preliminary characterization was conducted through thermogravimetric analysis (TGA) to establish the degradation temperature; infrared (IR) spectroscopy to gain information on the main constituents; and rotational rheology to understand the thermal stability over time at a given temperature. Samples to be tested through dynamic mechanical analysis (DMA) were obtained by a 3D printing machine at a temperature of 210 °C, by setting, as design parameters: an infill density equal to 70%, a layer thickness of 0.19 mm and a linear pattern.

Thermogravimetric measurements were performed using a Q500 TGA (TA Instruments, NewCastle, DE, USA). Tests were conducted by heating a piece of material (about 10 mg) at a rate of 20 °C/min from room temperature to 600 °C in an inert nitrogen atmosphere (purge gas flow of 50 mL/min).

Infrared spectroscopy was conducted in attenuated total reflectance (ATR) mode, using a spectrometer (model Spectrum 65 FT IR), produced by Perkin Elmer (Waltham, MA, USA), endowed with a diamond crystal. A wavenumber range of 400–4000 cm^{-1}, a resolution of 4 cm^{-1} and a number of scans equal to 16 were adopted. This analysis was carried out on surfaces of as-received filaments.

To characterize viscoelastic properties of molten polymers, and verify the thermal stability of materials, time sweep tests were performed through a rotational rheometer (model ARES), produced by TA Instruments (New Castle, DE, USA). Parallel plates 25 mm in diameter and a gap of 1 mm were adopted for the investigation. Materials were subjected to small-amplitude oscillations at a frequency of 1 rad/s and a strain amplitude of 1% for 10 min at 210 °C in an inert nitrogen atmosphere. Both samples were tested after being dried at 80 °C in a vacuum oven for 10 h and without being dried. The rheological characterization was then continued with oscillatory tests performed at a variable frequency in a range from 100 to 0.01 rad/s, keeping the amplitude equal to 1%, and the temperature at 210 °C, on dried samples.

The dynamic mechanical properties (DMA) of PLA filaments were examined using a Triton Technology Ltd. (Leicestershire, UK) instrument (model Tritec 2000). Rectangular specimens of 2 mm × 5 mm × 25 mm were investigated in single cantilever mode at frequencies of 1 Hz from room temperature to 80 °C, at a heating rate of 2 °C/min.

3. Results

3.1. Infrared Spectroscopy

The results of IR spectroscopy are shown in Figure 1, for pristine (black curve) and recycled (red curve) PLA filaments. The absorbance values were normalized in relation to an internal standard for the PLA polymer (1455 cm^{-1} peak [11]).

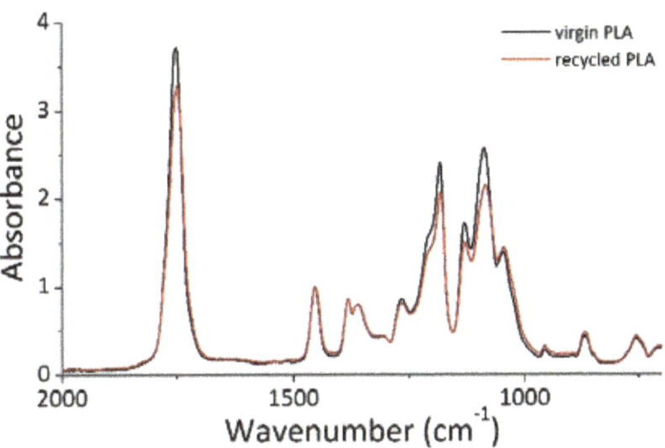

Figure 1. ATR spectra of PLA-based materials: virgin PLA (black line) and recycled PLA (red curve). Normalized peak at 1455 cm^{-1}.

PLA's characteristic peaks of the occurrence of oxidation and decomposition phenomena could be identified in both filaments: (i) 1750 cm^{-1}, linked to carbonyl (C=O) stretching; (ii) 1183 cm^{-1} and 1085 cm^{-1}, attributed to the asymmetric vibration of the ester group (C-O-C) [12]. In the case of the recycled material compared to the virgin one, a small reduction in the intensity of the correspondence of typical absorption bands of the PLA polymer was verified. This was interpreted a sign of weak thermal degradation of the PLA polymer [13] caused by exposure of the material to high temperatures during melt reprocessing and recycling operations. However, differences between spectra of pristine and processed PLA were not so substantial by displaying similar patterns of absorption. Therefore, it was excluded that a significant change in molecular structure took place upon reprocessing [14].

3.2. Thermogravimetric Analysis

During heating in the thermogravimetric analysis (Figure 2), one step of PLA degradation was shown in both samples. This trend was due to the loss of the ester group [15] that started at about 310 °C.

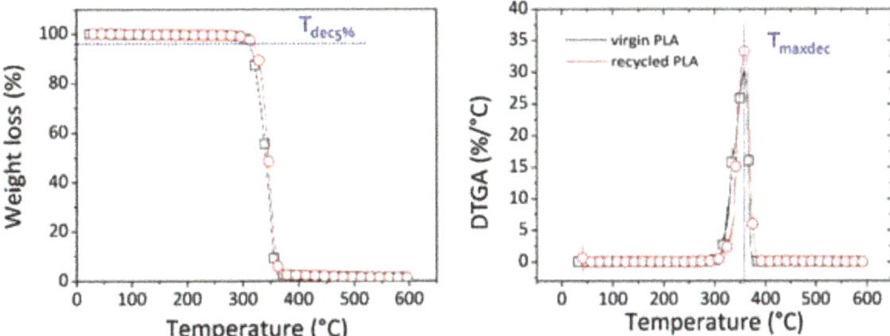

Figure 2. TGA thermograms of virgin and recycled filaments and their respective DTG (derivative thermogravimetric) curves.

In Table 1, the initial decomposition temperature ($T_{dec5\%}$), the temperature corresponding to the maximum rate of degradation (T_{maxdec}) and the final residue at 600 °C are reported for two analyzed samples.

Table 1. Initial decomposition temperature ($T_{dec5\%}$), temperature of maximum rate of decomposition (T_{maxdec}) and residue at a temperature of 600 °C.

	$T_{dec5\%}$	T_{maxdec}	Residue
Virgin PLA	315 °C	357 °C	1.5%
Recycled PLA	319 °C	360 °C	1.4%

For both polymers, the temperature at 5% of weight loss was recorded at around 320 °C, the maximum rate of thermal degradation was found at around 360 °C and a final residue of 1.5% remained. These data were compared with the TGA analysis on pellets of the same polymer (Ingeo 4043D by Naturework) reported in the literature. According to the study of da Cruz Faria et al. [16], neat PLA (Ingeo 4043D by Naturework) showed an onset temperature of mass loss of 300 °C, a temperature of the maximum mass loss rate Tpeak of 348 °C and a final residue of about 1% after 600 °C. The author concluded that this grade of commercial PLA was endowed with a high degree of purity, with TGA data in the expected range [13,17,18].

3.3. Rotational Rheology

Figure 3 depicts the experimental results of time and frequency sweep tests in terms of complex viscosity (Pa·s) over time (10 min) (Figure 3a), and complex viscosity (Pa·s) over angular frequency (Figure 3b), for the investigated systems.

Similar to other polyester polymers, PLA is sensitive to hydrolysis under melt processing conditions in the presence of a small amount of water [19]. In fact, as verified from the data in Figure 3a, a reduction in the rheological signal was attested, up to 30%, in 10 min when non-dried materials were analyzed. On the contrary, stabilization of the complex viscosity over time at a temperature of 210 °C was obtained through sample drying.

At a temperature of 210 °C, the complex viscosity trend as a function of the angular frequency (Figure 3b) displayed a Newtonian behavior for both polymers at low angular frequencies up to 10 rad/s, during which the value remained almost constant around 2–3×10^3 Pa s. Then, a shear thinning behavior followed for frequencies over 10 rad/s, during which the complex viscosity decreased.

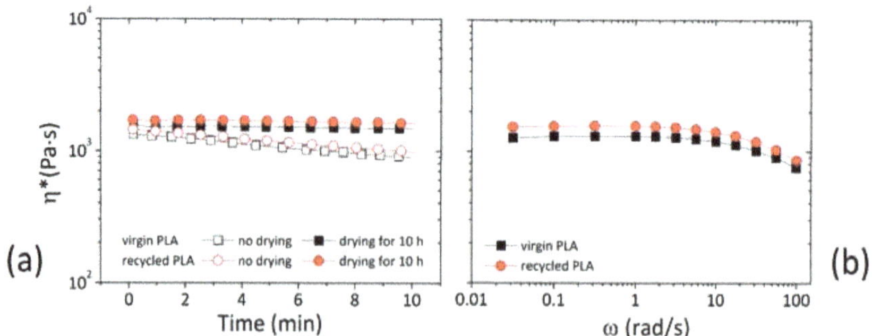

Figure 3. (a) Complex viscosity as a function of time for dried and non-dried specimens; (b) complex viscosity as a function of angular frequency for dried specimens.

In general, the main problem of PLA recycling is the polymer thermal instability attributed to the presence of moisture, lactic acid residues and metal catalysis that favor thermal degradation by leading to a reduction in the molecular weight and, consequently, the material viscosity [20]. In this case, no strong difference was observed between the complex viscosities of virgin and recycled PLA. This result highlights the negligible effects of recycling on the physico-chemical properties of resin.

3.4. Dynamic Mechanical Analysis (DMA)

The experimental results of DMA on 3D printed parts made from virgin and recycled filaments are shown in Figure 4 in terms of the storage modulus (E') (Figure 4a) and dissipation factor (tan delta) (Figure 4b) as a function of the temperature. In both cases, storage modulus curves underwent the typical sharp decline from 30 °C to 80 °C, related to glassy-to-rubbery state transition [12]. This behavior was usually attributed to an increase in the molecular mobility of polymer chains as the temperature was increased [21]. The dissipation factor (tan delta) provided the damping ability of the overall system, and the value of the glass transition temperature (Tg) corresponded to its maximum point [12]. According to the work of Pillin et al. [20], PLA possessed a glass temperature of 66.2 °C that decreased to 56.5 °C after seven injection molding processes. The authors concluded that a higher mobility of polymer chains attributed to a chain scission occurred during the reprocessing and recycling. From our data, E' and tan delta curves corresponding to virgin (black square points) and recycled samples (red circle points) roughly overlapped with almost comparable values across the entire temperature range. The value of the glass transition temperature in both cases was around 62 °C. Therefore, also through DMA analysis, it was confirmed that if a possible polymer degradation occurred during recycling, this phenomenon was mild and did not involve an alteration in the thermo-mechanical properties.

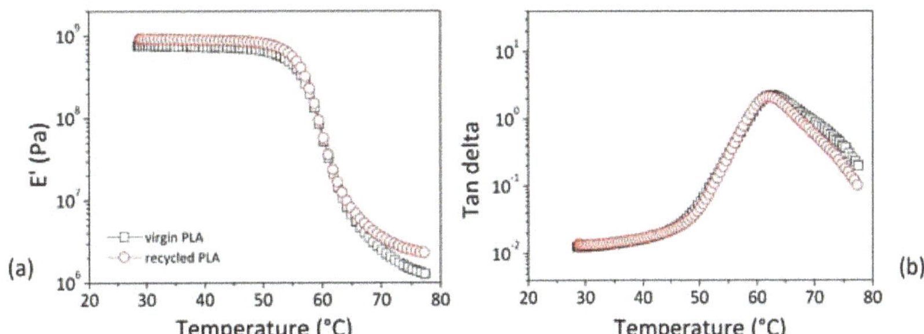

Figure 4. (**a**) Storage modulus (E') and (**b**) dissipation factor (tan delta) against temperature for 3D printed parts made from virgin and recycled filaments.

4. Conclusions

This was a preliminary study devoted to understanding the applicability of recycled matrices instead of virgin polymers for the 3D printing process. From the data, despite a small reduction in the ATR spectra, in correspondence with the PLA characteristic peaks of thermal degradation, no substantial differences could be highlighted in terms of thermal degradation, rheological behavior and thermo-mechanical properties. In fact, for both materials, the initial degradation temperature was measured at around 310 °C, the glass transition temperature was around 62 °C and the stability of the complex viscosity over time was achieved through sample pre-drying. The complex viscosity as a function of the angular frequency and the storage modulus of 3D printed parts made from recycled matrices were very comparable with those of the virgin matrices.

Supplementary Materials: The following are available online at https://www.mdpi.com/article/10.3390/IOCPS2021-11209/s1.

Author Contributions: Conceptualization, A.P. and D.A.; validation, G.C.; data curation, S.A. and M.Z.; writing—original draft preparation, A.P.; writing—review and editing, S.A. and G.C.; supervision, D.A. All authors have read and agreed to the published version of the manuscript.

Funding: This research received no external funding.

Institutional Review Board Statement: Not applicable.

Informed Consent Statement: Not applicable.

Data Availability Statement: The data presented in this study are available on request from the corresponding author.

Acknowledgments: A. Patti wishes to thank the Italian Ministry of Education, Universities and Research (MIUR) in the framework of Action 1.2 "Researcher Mobility" of The Axis I of PON R&I 2014–2020 under the call "AIM- Attrazione e Mobilità Internazionale". S. Acierno and G. Cicala acknowledge the support of the Italian Ministry of University, project PRIN 2017, 20179SWLKA "Multiple Advanced Materials Manufactured by Additive technologies (MAMMA)".

Conflicts of Interest: The authors declare no conflict of interest.

References

1. Patti, A.; Nele, L.; Zarrelli, M.; Graziosi, L.; Acierno, D. A Comparative Analysis on the Processing Aspects of Basalt and Glass Fibers Reinforced Composites. *Fibers Polym.* **2021**, *22*, 1449–1459. [CrossRef]
2. Pinto Costa, J.D.A.; Rocha-santos, T.; Duarte, A.C. The Environmental Impacts of Plastics and Micro-Plastics Use, Waste and Pollution: EU and National Measures. Available online: https://www.europarl.europa.eu/thinktank/en/document/IPOL_STU(2020)658279 (accessed on 27 September 2021).
3. Balart, R.; Montanes, N.; Dominici, F.; Boronat, T.; Torres-Giner, S. Environmentally friendly polymers and polymer composites. *Materials* **2020**, *13*, 4892. [CrossRef] [PubMed]
4. Patti, A.; Cicala, G.; Tosto, C.; Saitta, L.; Acierno, D. Characterization of 3D Printed Highly Filled Composite: Structure, Thermal Diffusivity and Dynamic-Mechanical Analysis. *Chem. Eng. Trans.* **2021**, *86*, 1537–1542. [CrossRef]
5. Moetazedian, A.; Gleadall, A.; Han, X.; Ekinci, A.; Mele, E.; Silberschmidt, V.V. Mechanical performance of 3D printed polylactide during degradation. *Addit. Manuf.* **2021**, *38*, 101764. [CrossRef]
6. Berezina, N.; Martelli, S.M. Bio-based Polymers and Materials. In *Renewable Resources for Biorefineries*; Royal Society of Chemistry: London, UK, 2014; pp. 1–28.
7. Farah, S.; Anderson, D.G.; Langer, R. Physical and mechanical properties of PLA, and their functions in widespread applications—A comprehensive review. *Adv. Drug Deliv. Rev.* **2016**, *107*, 367–392. [CrossRef] [PubMed]
8. Spierling, S.; Venkatachalam, V.; Mudersbach, M.; Becker, N.; Herrmann, C.; Endres, H.J. End-of-Life Options for Bio-Based Plastics in a Circular Economy—Status Quo and Potential from a Life Cycle Assessment Perspective. *Resources* **2020**, *9*, 90. [CrossRef]
9. McKeown, P.; Jones, M.D. The Chemical Recycling of PLA: A Review. *Sustain. Chem.* **2020**, *1*, 1–22. [CrossRef]
10. Badia, J.D.; Ribes-Greus, A. Mechanical recycling of polylactide, upgrading trends and combination of valorization techniques. *Eur. Polym. J.* **2016**, *84*, 22–39. [CrossRef]
11. Cuadri, A.A.; Martín-Alfonso, J.E. Thermal, thermo-oxidative and thermomechanical degradation of PLA: A comparative study based on rheological, chemical and thermal properties. *Polym. Degrad. Stab.* **2018**, *150*, 37–45. [CrossRef]
12. Patti, A.; Acierno, D.; Latteri, A.; Tosto, C.; Pergolizzi, E.; Recca, G.; Cristaudo, M.; Cicala, G. Influence of the processing conditions on the mechanical performance of sustainable bio-based PLA compounds. *Polymers* **2020**, *12*, 2197. [CrossRef] [PubMed]
13. Rasselet, D.; Ruellan, A.; Guinault, A.; Miquelard-Garnier, G.; Sollogoub, C.; Fayolle, B. Oxidative degradation of polylactide (PLA) and its effects on physical and mechanical properties. *Eur. Polym. J.* **2014**, *50*, 109–116. [CrossRef]
14. Signori, F.; Coltelli, M.B.; Bronco, S. Thermal degradation of poly(lactic acid) (PLA) and poly(butylene adipate-co-terephthalate) (PBAT) and their blends upon melt processing. *Polym. Degrad. Stab.* **2009**, *94*, 74–82. [CrossRef]
15. Chrysafi, I.; Ainali, N.M.; Bikiaris, D.N. Thermal Degradation Mechanism and Decomposition Kinetic Studies of Poly(Lactic Acid) and Its Copolymers with Poly(Hexylene Succinate). *Polymers* **2021**, *13*, 1365. [CrossRef] [PubMed]
16. da Cruz Faria, É.; Dias, M.L.; Ferreira, L.M.; Tavares, M.I.B. Crystallization behavior of zinc oxide/poly(lactic acid) nanocomposites. *J. Therm. Anal. Calorim.* **2021**, *146*, 1483–1490. [CrossRef]
17. Xiang, S.; Feng, L.; Bian, X.; Li, G.; Chen, X. Evaluation of PLA content in PLA/PBAT blends using TGA. *Polym. Test.* **2020**, *81*, 106211. [CrossRef]
18. Ainali, N.M.; Tarani, E.; Zamboulis, A.; Črešnar, K.P.; Zemljič, L.F.; Chrissafis, K.; Lambropoulou, D.A.; Bikiaris, D.N. Thermal Stability and Decomposition Mechanism of PLA Nanocomposites with Kraft Lignin and Tannin. *Polymers* **2021**, *13*, 2818. [CrossRef] [PubMed]
19. Van Den Oever, M.J.A.; Beck, B.; Müssig, J. Agrofibre reinforced poly(lactic acid) composites: Effect of moisture on degradation and mechanical properties. *Compos. Part A Appl. Sci. Manuf.* **2010**, *41*, 1628–1635. [CrossRef]

20. Pillin, I.; Montrelay, N.; Bourmaud, A.; Grohens, Y. Effect of thermo-mechanical cycles on the physico-chemical properties of poly(lactic acid). *Polym. Degrad. Stab.* **2008**, *93*, 321–328. [CrossRef]
21. Aw, Y.Y.; Yeoh, C.K.; Idris, M.A.; Teh, P.L.; Hamzah, K.A.; Sazali, S.A. Effect of Printing Parameters on Tensile, Dynamic Mechanical, and Thermoelectric Properties of FDM 3D Printed CABS/ZnO Composites. *Materials* **2018**, *11*, 466. [CrossRef] [PubMed]

materials
proceedings

MDPI

Proceeding Paper

Study and Characterization of Phase Change Material-Recycled Paperboard Composite for Thermoregulated Packaging Applications [†]

Tejashree Amberkar * and Prakash Mahanwar

Department of Polymer and Surface Engineering, Institute of Chemical Technology, Mumbai 400019, India; pa.mahanawar@ictmumbai.edu.in
* Correspondence: tejuamberkar@yahoo.co.in
† Presented at the 2nd International Online Conference on Polymer Science—Polymers and Nanotechnology for Industry 4.0, 1–15 November 2021; Available online: https://iocps2021.sciforum.net/.

Abstract: Beeswax is a bioderived phase change material (PCM), with a phase change temperature of around 60 °C, which is suitable for thermoregulating hot served food packages. Beeswax should be shape stabilized for thermoregulation purposes. This paper has used recycled paperboard as the matrix for shape stabilizing beeswax. Beeswax showed melting enthalpy of 216.09 J/g and melting enthalpy of composite with beeswax content 45% was 102.51 J/g. Thermal conductivity of beeswax and composite with 45% beeswax was calculated with the T-history method as 0.285 and 0.157, respectively. To address concerns, such as toxicity, environment-friendly nature and recycling, beeswax-recycled paperboard composite should be considered as a promising candidate.

Keywords: phase change material; thermal energy storage; beeswax; latent heat; packaging

Citation: Amberkar, T.; Mahanwar, P. Study and Characterization of Phase Change Material-Recycled Paperboard Composite for Thermoregulated Packaging Applications. *Mater. Proc.* **2021**, *7*, 17. https://doi.org/10.3390/IOCPS2021-11208

Academic Editor: Andreia F. Sousa

Published: 20 October 2021

Publisher's Note: MDPI stays neutral with regard to jurisdictional claims in published maps and institutional affiliations.

1. Introduction

Hot served food items are maintained around a temperature of 60 °C. Delay in delivery of food packages caused due to interventions in supply chain activities reduces the temperature of food products. Such products can be complemented with phase change material (PCM) pouches which assure temperature maintenance. PCM absorbs a large amount of heat at its phase transition temperature and releases heat to a cooler environment to establish a hot food temperature. Generally, PCMs used for packaging applications undergo a solid to liquid phase transition. These PCMs require an additional envelope to avoid leakage. The secondary envelopes of commercially available PCMs come in the form of rigid slabs or flexible pouches. These envelopes are made from non-biodegradable plastic. Many research groups have used PCM in delivering temperature-sensitive food products. Wang et al. used [1] microencapsulated PCM stored in a zip lock bag to transport meat package from a plastic crisper while maintaining a temperature around 4 °C. The microencapsulated PCM was in powder form. It required a zip lock bag of plastic crisper size to maintain a uniform weight of PCM along walls. This arrangement increased temperature retention time from 3 min to 30 min. Energy consumption of reefer transported food items was reduced by combining PCM into the insulation walls of vehicles [2–4]. PCM encapsulated into metal cold plates ensured uniform distribution of PCM along the reefer's walls. These plates were charged at the phase transition temperature of PCM by passing refrigerant through it. These cold plates were designed in fixed dimensions by large scale manufacturers. Such composite structures are suitable for transporting food items without refrigeration in 6–9 h. These heavy plates are useful only for big containers. Improvement in insulation performance increases temperature maintenance time. The combination of the vacuum insulation panel and PCM panel in a small container package can maintain the temperature in the range of 2–8 °C for 72 h without any refrigeration system [5]. Even hot beverage cups maintain sipping temperature for a longer duration with the combination of PCM layer and insulation layer [6].

In this paper, a PCM composite was prepared with reusable, biodegradable materials. This composite structure is comprised of recycled paper and beeswax. The porous structure of paper helps in shape-stabilizing the beeswax's phase transformation. The secondary plastic envelope is not required while using the prepared beeswax paper composite. Commercial PCM slabs are manufactured in a limited number of sizes as decided by the manufacturer. These sizes can be a bulkier option for incorporation into small packages. On the other hand, the designed composite can be transformed into the required shape as per package dimensions. This will help in minimizing package weight. Overall, the simple design and method of preparation of beeswax composite make it suitable for low cost commercial applications. The prepared PCM composite assures the temperature maintenance of food items for a longer time in the food delivery process while utilizing less energy.

2. Materials and Methods

2.1. Materials

Beeswax (Anaha™, Panaji, India) was procured. Commercially available paper board obtained from the local market (Mumbai, India). Tween 80 (S D Fine-Chem limited, Mumbai, India) was obtained. Sodium benzoate (Shree Lakshmi chemicals, Banglore, India) was purchased. DI water was used for experimental work. A four-channel temperature data logger was assembled and programmed with Arduino UNO based microcontroller (Adiy™, Rajguru electronics private limited, Mumbai, India). The data logger measures temperature with a precision of 0.01 °C. Two cartons of dimensions 19 cm × 13 cm × 25 cm were used. An insulation sheet of expanded polystyrene was procured from the local market (Mumbai, India). Two smaller cartons of dimensions 16 cm × 10 cm × 22.5 cm were purchased from the local market (Mumbai, India).

2.2. Method of Preparation

Paperboards were cut into pieces, and boiled water was added to prepare a 5% solution. The solution was kept overnight to obtain the paper pulp slurry. The slurry was mechanically ground. Beeswax was melted at 70 °C. The boiled water was added to the molten wax for preparing a 5% solution. The emulsifier tween 80 and the antimicrobial agent sodium benzoate were added to the solution in the quantity one pecent and two pecent respectively. The mixture was stirred for 30 min at 70 °C on a hot plate magnetic stirrer (Bexco, Haryana, India). Prepared paper pulp added to beeswax dispersion and stirred for 30 min at 70 °C. The quantity of pulp varied to prepare dispersion with 0%, 20%, 40%, 60%, and 80% beeswax concentration. Prepared dispersion was filtered with filter paper of pore size 8 to 10 microns. The suspension was filtered by vacuum filtration with a Buchner funnel. The material that remained above filter paper was peeled off and kept in a silicone mold of 9 cm × 9 cm × 0.2 cm size. It was then hot pressed with fabric covering. This process was useful in fabricating sheets of required dimensions. The prepared sheets were further dried for seven days.

2.3. Characterization Techniques

2.3.1. Leakage Test

The highest concentration of beeswax showing no leakage was determined with the leakage test. The leakage can be determined by quantitative and qualitative methods. In the qualitative method, the circular sheet samples were heated to 70 °C. The leakage of molten beeswax observed can be observed visually. It gave the twenty percent range of beeswax concentration with the lowest leakage level. In this range, three composites were prepared by increasing beeswax concentration in steps of 5%. These samples will help to determine the highest quantity of beeswax that can be incorporated into the paper. The quantitative leakage test was performed for all samples. The quantitative analysis is based on the gravimetric principle. One to two grams of composite material and a small tissue paper was filled in a test tube. The test tube was heated to 70 °C. The weight of paper before and after heating was tested. If the weight increases, it indicates leakage from the composite.

2.3.2. Differential Scanning Calorimetry

About 5 mg of sample was weighed on electronic weighing balance (AB204, Mettler Toledo, Tokyo, Japan) used for the analysis. The machine can weigh ±0.1 mg precisely. Differential scanning calorimeter (DSC 3, Mettler Toledo, Tokyo, Japan) was used. To remove the thermal history of the sample, the sample was heated from room temperature to 200 °C at a heating rate of 10 °C/min under a nitrogen atmosphere. The sample was held at 200 °C for 2 min. The melting characteristics of the sample were analyzed in the next scan. The scanning was performed from 0 °C to 200 °C at a heating rate of 10 °C/min under a nitrogen atmosphere.

2.3.3. T-History Method

The experimental setup consisted of two identical test tubes consisting equal amount of glycerine (reference) and sample PCM, heating assembly and temperature sensors. The sample and reference were heated to 70 °C and allowed to cool down to room temperature. The temperature drops in the reference and sample, in the temperature range 70 °C to 27 °C, were recorded. These values were used to plot temperature-time curves of the PCM sample and reference.

The specific heat and thermal conductivity of the solid PCM are determined using the following formula:

$$c_p = \frac{m_R \times c_{p,R} + m_t \times c_{p,t}}{m_p} \times \frac{A_p}{A_R} - \frac{m_t \times c_{p,t}}{m_p}$$

where m_R is mass of reference material, m_t is mass of tube material, m_p is mass of PCM, c_p is the specific heat capacity of PCM sample, $c_{p,R}$ is the specific heat capacity of reference material, $c_{p,t}$ is the specific heat capacity of tube material, A_p is the area under the cooling curve of PCM and A_R is the area under the cooling curve reference material.

The thermal conductivity of the solid PCM was calculated using the formula:

$$k_p = [1 + \frac{c_p(T_m - T_\infty)}{H_m}]/4\left[\frac{t_f(T_m - T_\infty)}{R^2 \rho_p H_m}\right]$$

where k_p is the effective thermal conductivity of the PCM, c_p is the specific heat of the PCM, ρ_p is the density of the PCM, R is the radius of the test tube, H_m is the heat of fusion of PCM as obtained from the differential scanning calorimetry (DSC) results and T_m & T_∞ is the temperature of melting and atmosphere, respectively. The time of solidification of the molten PCM is denoted by t_f.

2.3.4. Morphological Analysis

The morphology of samples was observed with the scanning electron microscope (SEM). FEI Quanta 200 SEM model was used for studying morphology.

2.3.5. Heat Release Performance in Carton

Heat release performance of PCM composite sheets was compared with a control sample. The test was performed in a room with an ambient temperature of 29 °C. The arrangement of the test components is given in Figure 1. In the PCM-recycled paper sample, PCM composite sheets were placed at six faces of the outer carton. Each side of the carton is surrounded by 15 gm of PCM composite sheet layers. In six faces, the carton had 60 gm of PCM. An insulation sheet was placed beside each PCM composite sheet. A smaller sized inner carton was placed in this assembly. A glass containing 50 mL of boiled water was placed inside the inner carton. A control sample was prepared with the same procedure without PCM sheets. The temperature of water in the center of both cartons and ambient temperature was measured with the temperature data logger. The data logger was sealed inside the carton for measuring the temperature of water without heat loss. The data logger was connected to a laptop which acts as a data acquisition unit. It records the temperature in 30 s intervals. The temperature change in the water of both samples was measured simultaneously with the data logger.

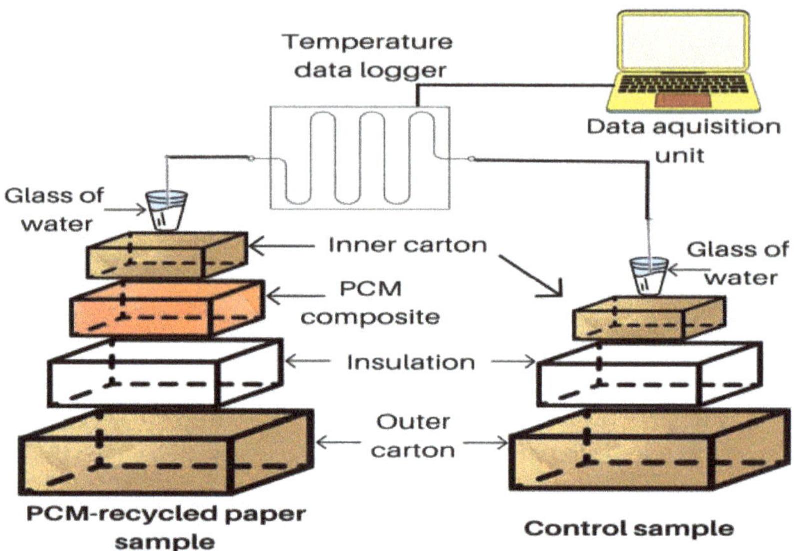

Figure 1. Assembly for heat release performance testing.

3. Results and Discussion

With the increasing content of beeswax, the enthalpy of phase transition increases. This implies that an increase in thermoregulation capacity is possible with high beeswax loading in the recycled paper composite. However, the limit of maximum beeswax incorporation in the recycled paper composite is constrained by leakage. The composite samples containing 60% and 80% beeswax concentration showed visual leakage at 70 °C in the qualitative test. The leakage can be observed in Figure 2. The lower concentration of fibers in these samples could not hold molten beeswax. The qualitative test proved the 40–60% range as the optimum beeswax concentration level for the composite. For determining the exact amount of beeswax concentration, the leakage test was performed on composite sheets with 45%, 50% and 55% beeswax concentration. These samples were made following the previously mentioned method. The results of the qualitative leakage test on the backside of the paper can be seen in Figure 2. The pure beeswax sample was fully molten. Samples above 45% beeswax concentration showed leakage. Only 45% of beeswax composite did not show leakage.

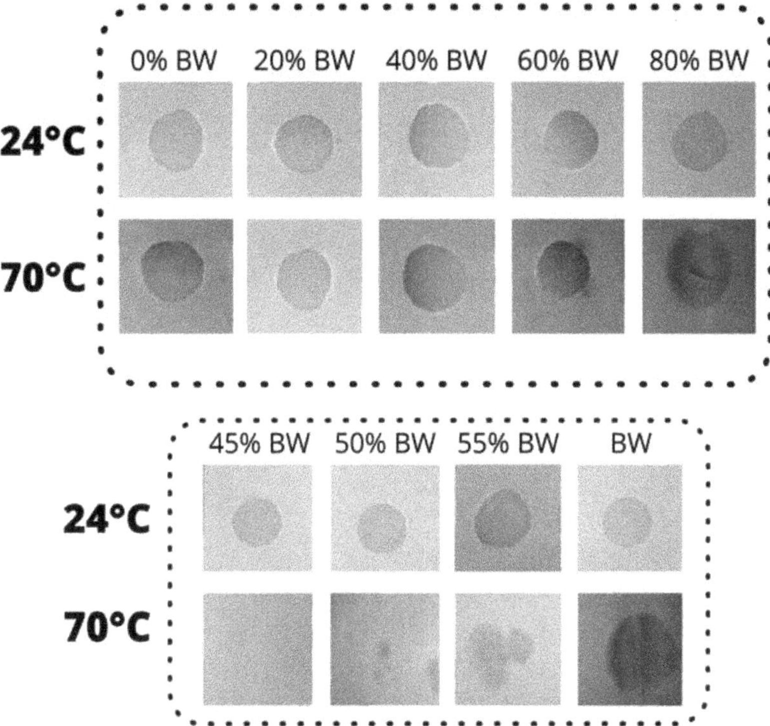

Figure 2. Leakage test of beeswax composites.

The quantitative leakage test was performed to confirm the qualitative leakage test's results. Samples with 20% and 40% beeswax concentration showed negligible leakage. Samples with 60% and 80% beeswax concentration showed a 12.66% and 20.38% increase in paper weight, respectively. The leakage test indicated that samples with 60 and 80% beeswax content showed severe leakage which is unsuitable for practical applications. Pores of paper cannot prevent leakage of PCM above its melting point. Leakage tests showed a 3.83% and 9.25% weight increase for 50% and 55% beeswax containing composites. Negligible leakage was obtained with 45% beeswax-recycled paper composite. Both qualitative and quantitative leakage tests confirmed 45% beeswax concentration as optimum. Thus, 45% beeswax is considered as the highest concentration that can be incorporated in paperboard with the mentioned technique.

The DSC thermograms of beeswax, beeswax composite containing 45% beeswax concentration, and paper are given in Figure 3. Beeswax is a bio-based material composed of long-chain acids and esters [7]. The long-chain compounds require a large quantity of heat for the melting process. This amount can be measured with DSC. Beeswax showed a melting enthalpy of 216.09 J/g. When micrograins of beeswax are enclosed in the porous paper, its crystalline structure is modified. The thermal characteristics change due to structural variation. Thus, the melting enthalpy of composite with 45% beeswax content was reduced to 102.51 J/g. A smaller quantity of latent heat was required for phase transition of modified crystalline structure, while a higher amount of latent heat was needed to transform the crystalline phase. The structural change also affected phase transition temperature. The beeswax and its composite underwent phase transition at 59.92 °C and 61.11 °C, respectively. The cellulosic paper did not undergo any structural change in beeswax's phase change temperature range.

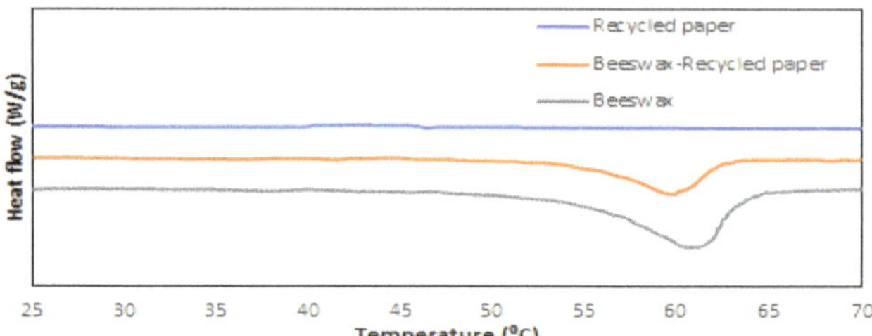

Figure 3. DSC thermograms for melting process.

A small quantity of samples was used for DSC analysis to give accurate results of thermal properties. However, practical applications use a large amount of material. The thermal properties could vary due to the non-homogeneity of samples. Thus, thermal tests performed using a higher amount of sample will be more useful. The T-history method is advantageous in this respect. It measures the thermal performance of larger sample weights. In this test, the temperature-time profile of the sample is compared with reference material for determining thermal characteristics. As shown in the temperature-time curve in Figure 4, beeswax has a larger area under the curve depicting the highest heat storage capacity. With decreasing beeswax content, the area under the temperature-time curve reduces. This means that a lower amount of heat is stored when paper content in the composite is higher. The thermal conductivity of beeswax and composite samples with 45%, 40% and 20% beeswax was calculated as 0.285 W/mK, 0.157 W/mK, 0.141 W/mK, and 0.119 W/mK, respectively. These values are comparable to observed values in the literature [8]. As the beeswax concentration decreases, the conductivity decreases. The beeswax is filled inside the pores of paper which otherwise is filled with air. That is why the increase in beeswax concentration increases the thermal conductivity of composite structures.

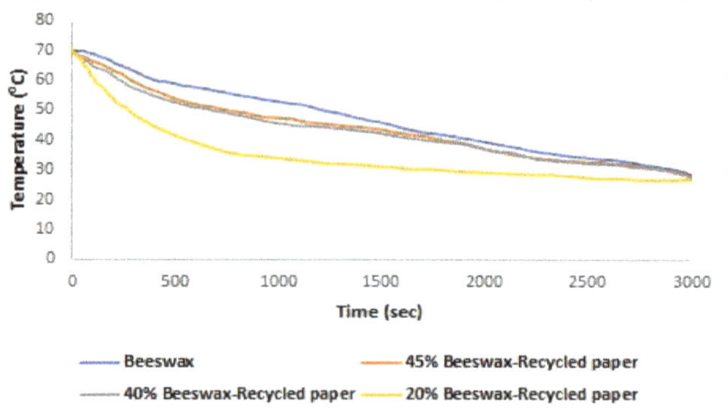

Figure 4. Temperature-time curve of beeswax and beeswax composite.

SEM analysis of the recycled paper is represented in Figure 5a. The morphology of recycled paper consisted of pores. The morphology of composite containing 45% beeswax concentration is shown in Figure 5b. The mesh structure of the paper was filled with beeswax. However, the fiber strands are visible. Beeswax was not on the surface of fibers. Otherwise, the wax would have leaked away in the molten state. So, it can be concluded that wax was only present in the pores of the paper. The porous structure helps in tethering leakage of molten beeswax. The formability of recycled paper-beeswax composite to different shapes and sizes combined with beeswax micrograins in paper pores is advantageous for thermal energy storage applications.

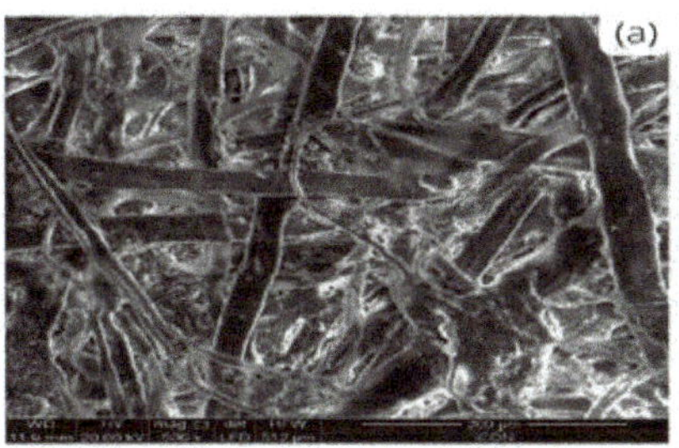
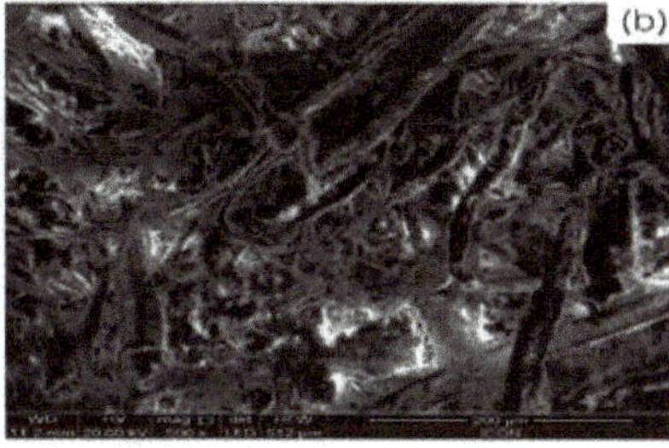

Figure 5. Morphology of (**a**) paper and (**b**) beeswax composite.

The effectivity of PCM composite sheets on the temperature of enclosed water was studied with the heat release performance test in a carton. Figure 6 gives the temperature-time profile of the water in the beeswax composite contained carton and control sample carton. The time of temperature measurement was the same for the samples. The starting points of temperature measurement for PCM composite samples and control samples were 76 °C and 70 °C, respectively. The initial temperature variation in control and composite samples was due to time lag in adjusting the data logger inside cartons. This point was considered as 0 s point. Hereon time measurement started. The Composite assembly took 6823 s for reaching ambient temperature. Control assembly took 6259 s to reach ambient temperature. An increase in time for reaching ambient temperature confirms the thermal energy storage phenomenon of PCM composite sheets. Initially, PCM temperature drops around 38 °C due to absorption of heat by PCM. This temperature was lesser than the control assembly. It was due to the higher thermal conductivity of beeswax composite sheets. Higher thermal conductivity also increases heat loss. Since the outer side of sheets was surrounded by insulation sheets, the heat travels only towards the glass of water in the inner direction. Thus, at 2362 s, the temperature started to increase with heat release from PCM. The temperature increased to a maximum of 43.8 °C. The time for reaching this temperature was 3265 s. It was expected that the maximum temperature of the water should be the phase transition temperature of composite, i.e., 59.92 °C. However, the distance traveled by heat from the PCM composite sheet to the center resulted in heat loss. Additionally, the edges of the inner carton hindered heat passage and contributed to heat loss. After reaching 43.8 °C, the temperature started decreasing and slowly reached ambient temperature. The composite assembly kept the inside temperature at a higher value for more time than the control assembly. This higher temperature inside the carton is from the heat release from PCM to the inside chamber. This phenomenon can be used to maintain a higher temperature in case of delayed delivery. The use of more weight of PCM composite sheets can increase temperature maintenance time. Reducing the distance between PCM composite sheets and food items can help in maintaining food near phase transition temperature of composite.

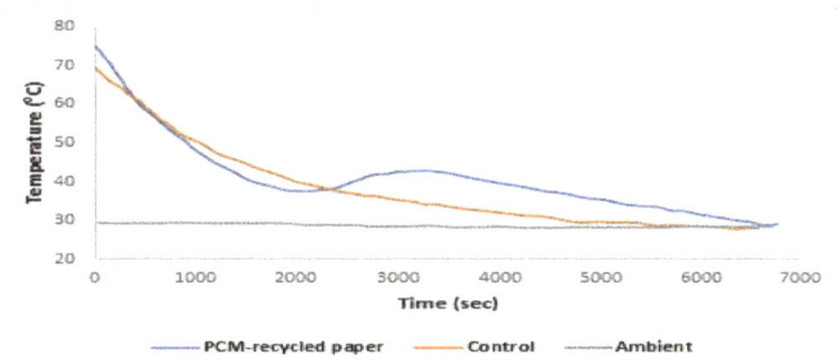

Figure 6. Temperature-time profile of control sample, composite sample and ambiance.

4. Conclusions

Optimization studies revealed that 45% is the maximum amount of beeswax that can be incorporated in the paper composite. The thermal conductivity of composites increases with an increase in beeswax concentration. The carton containing PCM sheets maintain the temperature at a higher level for a longer time than for carton without PCM. The phase transition temperature of the prepared composite is around 60 °C. If food items are prepared at once and stored in a container equipped with beeswax-recycled paper composite, it can maintain the temperature of food items for a longer time in the food delivery process while utilizing less energy.

Author Contributions: Writing—original draft preparation, T.A.; review and editing, P.M. All authors have read and agreed to the published version of the manuscript.

Funding: This work is funded by AICTE National Doctoral Fellowship Scheme sanctioned vide letter F No: 12-2/2019-U1 provided by the Ministry of Human Resource Development, Government of India.

Institutional Review Board Statement: Not applicable.

Informed Consent Statement: Not applicable.

Data Availability Statement: The data presented in this study are available on request from the corresponding author.

Acknowledgments: The authors acknowledge research facilities provided by Institute of Chemical Technology, Mumbai.

Conflicts of Interest: The authors declare no conflict of interest.

References

1. Wang, Y.; Zhang, Q.; Bian, W.; Ye, L.; Yang, X.; Song, X. Preservation of traditional Chinese pork balls supplemented with essential oil microemulsion in a phase-change material package. *J. Sci. Food Agric.* **2020**, *100*, 2288–2295. [CrossRef] [PubMed]
2. Zhang, S.; Zhang, X.; Xu, X.; Zhao, Y. Thermomechanical Analysis and Numerical Simulation of Storage Type Multi-Temperature Refrigerated Truck. *Int. J. Sci.* **2020**, *7*, 261–270.
3. Mousazade, A.; Rafee, R.; Valipour, M.S. Thermal performance of cold panels with phase change materials in a refrigerated truck. *Int. J. Refrig.* **2020**, *120*, 119–126. [CrossRef]
4. Radebe, T.B.; Huan, Z.; Baloyi, J. Simulation of eutectic plates in medium refrigerated transport. *J. Eng. Des. Technol.* **2021**, *19*, 62–80. [CrossRef]
5. Huang, L.; Piontek, U. Improving performance of cold-chain insulated container with phase change material: An experimental investigation. *Appl. Sci.* **2017**, *7*, 1288. [CrossRef]
6. Booska, R. Thermal Receptacle with Phase Change Material. U.S. Patent 10,595,654, filed 17 May 2018, and issued 24 March 2020,
7. Amberkar, T.; Mahanwar, P. Composite phase change material for improving thermal protection performance of insulated packaging container. *Int. J. Eng. Trends Technol.* **2022**, *70*, 59–64. [CrossRef]
8. Dinker, A.; Agarwal, M.; Agarwal, G.D. Preparation, characterization, and performance study of beeswax/expanded graphite composite as thermal storage material. *Exp. Heat Transf.* **2017**, *30*, 139–150. [CrossRef]

materials proceedings

Proceeding Paper

Current Alternatives for In-Can Preservation of Aqueous Paints: A Review †

Pieter Samyn *, Joey Bosmans and Patrick Cosemans

SIRRIS Smart Coating Application Lab, B-3590 Diepenbeek, Belgium; joey.bosmans@sirris.be (J.B.); patrick.cosemans@sirris.be (P.C.)
* Correspondence: pieter.samyn@sirris.be
† Presented at the 2nd International Online Conference on Polymer Science—Polymers and Nanotechnology for Industry 4.0, 1–15 November 2021; Available online: https://iocps2021.sciforum.net/.

Abstract: With the transition towards more sustainable paint formulations that are waterborne, the susceptibility to microbial contamination has to be better controlled to increase shelf life and functional lifetime. However, recent restrictions in European regulations on the use of biocides have put limitations on the concentrations for traditional systems providing either in-can or dry-film preservation. The commercial technologies for in-can preservation that are currently available are based on isothiazolines, such as 2-methyl-4-isothiazolin-3-one (MIT), 1,2-benzisothiazolin-3-one (BIT) and 5-chloro-2-methyl-4-isothiazolin-3-one (CMIT). At present, however, only a limited number of alternatives can be used and are reviewed in this presentation. Examples of non-sensitizing biocidal components for coatings include quaternary/cationic nitrogen amines, silver ions and zinc complexes. However, the use of the latter is not without risk to human health. Therefore, it is believed that disruptive methods will need to be implemented in parallel with more innovative bio-inspired solutions. In particular, the antimicrobial polymers, amino-acid-based systems and peptides have similar functions in nature and can offer antimicrobial activities. Additionally, cross-border solutions currently applied in food or cosmetics industries should be considered as examples that need to be further adapted for paint formulations. However, incorporation in paint formulations remains a challenge in view of the stabilization and rheology control needed for paint. This work's overview aims to provide different strategies and best evidence for future trends.

Keywords: paint; anti-microbial; biocide; preservation

Citation: Samyn, P.; Bosmans, J.; Cosemans, P. Current Alternatives for In-Can Preservation of Aqueous Paints: A Review. *Mater. Proc.* **2021**, *7*, 18. https://doi.org/10.3390/IOCPS2021-11245

Academic Editor: Ana María Díez-Pascual

Published: 30 October 2021

Publisher's Note: MDPI stays neutral with regard to jurisdictional claims in published maps and institutional affiliations.

1. Introduction

Waterborne paint formulations have been introduced in response to demands for more eco-friendly solutions with fewer or no chemical solvents and lower VOC emissions. In parallel, however, such aqueous environments are beneficial for the growth and survival of micro-organisms such as bacteria, fungi and yeast. They can get into paint via the raw materials or various contamination sources in the processing plant. In a particular study, microbial contamination of the paint with *Pseudomonas* as the predominant genus mainly occurred as a result of biofilm formation in the production equipment [1]. In another study, sufficient screening and appropriate selection of the raw materials was advised in order to reduce contamination [2]. The degradation of paints in the presence of micro-organisms can be noticed by a change in color, a penetrating odor, gas formation, reduced stability, a pH variation and a viscosity reduction. The quality loss of tainted paint finally results in product spoilage and time delay. Although the presence of bacteria can be controlled through better material selection and plant hygiene, they cannot be fully avoided. Therefore, in-can preservation (PT-6) is required to ensure a long in-pot lifetime, which is industrially expected to be at least 3 years.

The use of biocides for in-can preservation (PT-6), which are inherently toxic and potentially affect human health, has been subjected to more stringent legislation in recent years, following the Biocidal Products Regulation (updated March 2020). In the early years, organo-mercury compounds and formaldehyde biocides were banned because of carcinogenic effects. Following risk assessment studies, the diverse range and different potential of formaldehyde-condensate compounds in regard to formaldehyde gas were recognized and taken into account for preservatives. While some standards for paints and coatings (Green Seal GS-11) prohibit the release of free formaldehyde, others have restricted the emission of free formaldehyde at below 100 ppm. Isothiazolinone derivatives were introduced as formaldehyde-free alternatives for both in-can and dry-film preservation, including 1,2-benzisothiazolin-3-one (BIT), 2-methylisothiazol-3(2H)-one (MIT), 2-octyl-2H-isothiazol-3-one (OIT), 5-chloro-2-methyl-4-isothiazolin-3-one (CMIT) and 4,5-dichloro-2-n-octyl-3(2H)-isothiazolone (DCOIT) (Figure 1) [3]. Nevertheless, the observations of allergic skin reactions towards one specific type of isothiazoline, i.e., MIT, lead to its classification as skin sensitizer and a reduction in its allowable concentration below 15 ppm according to the Risk Assessment Committee (RAC) of the European Chemicals Agency (ECHA). In view of a harmonized classification, concerns rose on the use of other isothiazolines at low dosages (<15 ppm) that are insufficient in-can preservation. In parallel, the present options for alternative preservatives are limited through the Article 95 List, on which 52 active ingredients are listed and only 15 are compatible with paints and coatings [4]. Therefore, the availability of biocides for in-can preservation is restricted and puts high pressure on industrial applications.

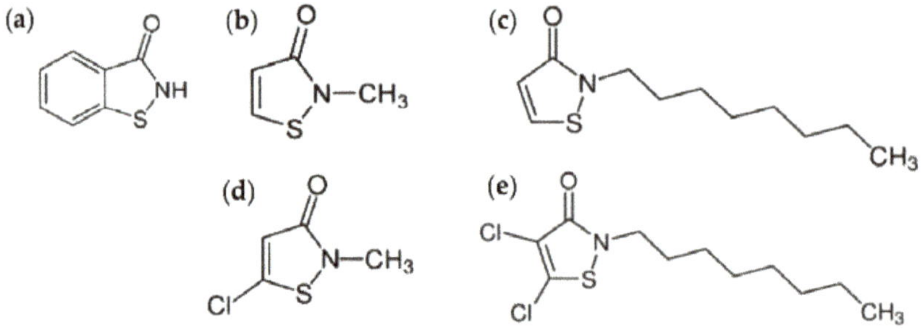

Figure 1. Isothiazolinones presently used in the industry for in-can preservation of paint: (**a**) BIT (CAS 2634-33-5), (**b**) MIT (CAS 2682-20-4), (**c**) OIT (CAS 26530-20-1), (**d**) CMIT (CAS 26172-55-4), (**e**) DCOIT (CAS 64359-81-5).

Present solutions for the short-term are limited, and long-term developments should take into account novel preservation systems, as reviewed in this contribution. This overview focusses on compositional aspects of waterborne paints or latex (e.g., acrylates) and does not detail additional measures that can be taken to enhance paint preservation, such as raw material screening, pasteurization, thermal treatments, plant hygiene and anti-septic packaging. Additionally, alternative systems such as high-pH paints and dispersible powder paints are not further discussed.

2. Current Industrial Trends in the Preservation of Waterborne Paint

2.1. Blending of Biocide Formulations

The limits for reducing concentrations of isothiazolines are set by the inhibitory concentrations of MIT and BIT required in order for them to function as efficient biocides (Table 1) [5]. For most single biocides, the minimum inhibitory concentration (MIC) is above the threshold value for selected laboratory strains, whereas even a higher concentration might be necessary in practice to mitigate wild strains [6]. It should be noticed that some biocides have deficiencies or "gaps" in their performance, e.g., the limited anti-fungal properties of BIT and MIT. By using biocide blends, a synergistic biocidal effect was observed with an impressive decrease in minimum inhibitory concentration, below regulatory limits. In parallel, the chances bacteria developing resistance against biocides significantly reduces when they are exposed to blends rather than single active ingredients [7]. Bacteria originating from biofilms are indeed known to have significantly increased tolerance towards common biocides used for in-can preservation, and the minimum inhibitory concentrations are indeed exceeded in paints. The biocide blending also allows one to combine biocides both short-term activity (e.g., O-formals or CMIT) and biocides that provide long-term protection (e.g., BIT). By selecting a combination of active ingredients within an optimum concentration ratio, compatibility can be created regarding properties such as pH and redox potential. The two most promising and versatile platforms for in-can preservation are based on a combination of MIT and BIT in a ratio of 1:1 (for pH ranges between 2 and 11), and a combination of CMIT and MIT in a ratio of 3:1 (for pH ranges below 8) [8]. By utilizing MIT/BIT and CMIT/MIT blends, new generations of in-can biocides were industrially developed that can outperform traditional biocides by a factor of two or more without cautionary labelling related to the European legislation [9]. In such formulations, CMIT and MIT can be used at concentrations below 15 ppm [10]. CMIT has been identified as a skin sensitizer at concentrations above 64 ppm, but that concentration has been avoided.

Table 1. Biocide activity (MIC values) of isothiazolines and their blends against laboratory-scale organism cultures (data summarized according to [5,6]).

Organism	Minimum Inhibitory Concentration (MIC) (ppm)				
	BIT	MIT	CMIT	MIT/BIT	CMIT/MIT
Escherichia coli (b)	25	17.5		10	9.0
Klebsiella pneumoniae (b)	25	20		15	9.0
Pseudomonas aeruginosa (b)	150	30	0.6	20	11.2
Pseudomonas putida (b)	60	12.5	0.2	10	
Pseudomonas stutzeri (b)	20	12.5		10	
Aspergillus niger (m)	100	750		50	9.0
Candida albicans (y)	200	75			9.0

b = bacteria, m = molds, y = yeast.

The blending of isothiazolines with 2-bromo-2-nitro-1,3-propanediol (bronopol), in particular, CMIT/MIT, was used to improve the efficiency of MIT at low concentrations. Dosages for those systems can vary with sample preparations, e.g., from 60 ppm bronopol + 10 ppm CMIT/MIT to 200 ppm bronopol + 33 ppm CMIT/MIT, depending on the severity of the contamination and the substrate. A concentration ratio of 6:1 bronopol to CMIT/MIT was recommended [11], and with bronopol, the dosage of CMIT/MIT can be reduced from 20 to 30 ppm to 7.5 to 15 ppm. Bronopol has been utilized to control bacteria that have developed tolerance or resistance against other biocides based on fomaldehyde or isothiazolines. Recently, the interest in bronopol was mainly raised for headspace preservation.

The combinations of BIT with pyrithione (PT) active agents, which are known as traditional fungicides, are finding increased use as co-biocides for in-can preservation of latexes. The pyrithiones interact with the microbial membranes as chelating agents and disrupt essential ion gradients. However, only specific Zn-pyrithiones (ZPT) or Na-pyrithiones (NPT) are applicable, as other cations result in strong coloration. The chelation complexes with Fe^{2+} or Cu^+ ions result in insoluble compounds and blue color for only small concentrations of ions present. The pass levels in antimicrobial testing for sample formulations of 2% BIT + 8% NPT and 2% BIT + 4% NPT were achieved at biocide concentrations of 0.10%. The biocide could be used up to a maximum concentration of 0.25% with an acceptable BIT level [12]. In particular, the BIT/ZPT blend is an efficient biocide against *Pseudomonas* bacteria that are more resistant against other preservatives (see Table 1).

Other types of biocide are added as co-preservatives to enhance the performance of CMIT/MIT and BIT at low dosages, such as 2,2-dibromo-3-nitrilopropionamide (DB-NPA), dichloro-2-n-octyl-4-isothiazoline-3-one (DCOIT), sorbate, sodium benzoate, ophenylphenol sodium salt and carbamate [13]. However, some of them are problematic due to various reasons related to the specific biocide/fungicide activity, color, hydrolysis, and high concentration requirements. Although covering a broad activity range against Gram-positive and Gram-negative bacteria, yeast, fungi and algae, DBNPA is known to have fast but brief activity. New micro-emulsion technologies need to be developed using DCOIT.

2.2. Metal-Based Additives

The antimicrobial properties of metal-oxide nanoparticles have been demonstrated in several fields, such as textiles, packaging, cosmetics and biomedicine. Paint formulators have mainly used silver ions or silver nanoparticles. The silver ions directly interact with the bacterial metabolism, preventing the conversion of nutrients into energy and inhibiting the survival, reproduction and colonization of bacteria. The silver additives can be based on silver phosphate glass active ingredients, and silver ions are effective against a broad spectrum of bacteria. However, in-can preservation with silver ions requires a high ion concentration to be efficient, and in consequence, is expensive, so the industrial application of this technique remains limited. Therefore, it is more attractive in use silver for dry-film protection, which also presents anti-fungal activity when used as additive in waterborne acrylic indoor paints [14].

Other metal oxides, such as ZnO, TiO_2, SiO_2 and MgO, have additional photocatalytic activity and can release reactive oxygen substances to kill bacteria under UV radiation. In particular, zinc oxide and other zinc complexes have been used as dry-film preservatives in exterior coatings to reduce fungi and algae growth. The addition of ZnO nanoparticles in a waterborne acrylic latex coating provided stable dispersions with enhanced physico-chemical, mechanical and anti-corrosive properties in parallel with antimicrobial resistance at a lower concentration than biocides in commercial paints [15]. Indeed, acrylic paints may benefit from ZnO nanoparticles through antimicrobial properties and low toxicity [16], and the active role of the oxide species may be further enhanced through the combination of ZnO nanoparticles in oxide/amine composites added into the acrylic paint [17]. Alternatively, the interior paints which are water-based acrylate dispersions with MgO nanoparticles provide antimicrobial properties due to the morphology of the nanoparticles (sharp edges) and the formation of reactive oxygen species that induce the peroxidation of lipids in the bacterial cells [18]. Functionalized SiO_2 mostly showed an effect against algal growth in dry paints and was also beneficial due to its non-leaching properties [19]. However, the future applications of metal nanoparticles must face toxicity risk assessments.

It may be wise to shift towards naturally-based nanoparticles with demonstrated antibacterial properties in waterborne polyurethane systems, including clay-based minerals. Halloysite nanotubes were formulated with encapsulated carvacrol and allowed sustained release and antibacterial activity over a long period [20].

3. Novel Bio-Based Trends in the Preservation of Waterborne Paint

The need to consider alternative preservation mechanisms instead of traditional isothiazolines is due to:

(i) All types of isothiazolines are inherent skin sensitizers, although at different concentrations. The classification of MIT as a skin sensitizer at a low concentration might set a precedent for other types, such as CMIT, CMIT/MIT and BIT, finally leading to homogenized legislation and reduced usage of all isothiazolines. The eventual scarcity in supply and banning of common isothiazolines threaten the paint industry.

(ii) Micro-organisms are developing increased tolerance against isothiazolines and need exposure to different classes of biocides.

Apart from current industrial approaches, more sustainable innovations for in-can preservation using bio-based compounds are to be expected. The development of novel mechanisms can be inspired through natural antimicrobial systems. Some natural polymeric materials exhibit antimicrobial activity, which can be considered for in-can preservation of paints. The presented solutions, however, are not yet common practice in the paint industry but are promising. Indeed, the development of novel in-can preservation will require appropriate development time and resources. The inspiration for natural preservatives can be found in other domains, such as cosmetics and food preservation, where commonly applied natural preservatives include plant extracts, chitosan or oligosaccharide derivatives, bacteriocins, bioactive peptides and essential oils.

3.1. Acids

Mild bio-based and/or biodegradable organic acids and combinations thereof are applied in cleaning, disinfection, food and personal care products. Their antibacterial mechanisms are based on combinations of effects, such as interacting with the bacterial cell membrane and acting as chelating agents to disrupt the sequestration of nutrients. Acidic conditions disrupt cell regulation at a general level and inhibit the fermentation processes, thereby preventing bacterial growth. The amphiphilic nature of these molecules and their combinations with acids or other surfactants often enable interactions with the cell membrane (lipid bilayer) and enhance cell permeability. Acids for in-can preservation are compatible with the low pHs of paint formulations, and their performances can be further boosted by other paint additives.

Lactic acid is a plant-based active ingredient against Gram-negative bacteria and is compatible with paint formulations that include a cationic surfactant system at have pHs below 4.5 [21]. The preservative effect of lactic acid can be attributed to the production of antimicrobial substances such as hydrogen peroxide and the promotion of an acidic environment. The concentration affinities of lactic acid for fifty strains were demonstrated, showing a clear inhibitory effect at low pHs that was maintained even at a neutral pH [22]. Zinc lactate is a zinc salt derived from lactic acid that also combines antimicrobial properties against bacteria and yeasts [23].

The use of other acids, such as caprylhydroxamic acid, benzoic acid, sorbic acid and their combinations with polyols such as glyverin, propylene glycol, propanediol and hexane diol, rely on the creation of multiple barriers against microbial growth and tunability against a broad spectrum of antimicrobial control [24].

3.2. Antimicrobial Polymers

The cationic polymers bearing positive charges and their assemblies may generally serve as favorable antimicrobial compounds [25]. The presence of quaternary ammonium groups in a polymer structure is a well-known example, as it also occurs in nature. Alternatives include polymer compounds with halogens, phosphor or sulphonates; organometalic polymers; and phenol and benzoic acid [26]. Cationic amines are available as industrial solutions; however, the incorporation of cationic biocides into an acrylic paint formulation is challenging and requires additional chemical modifications, as most water-based paints are anionically stabilized. As such, an antimicrobial waterborne polyurethane paint was developed via transformation into a quaternary ammonium salt emulsion [27]. The modified paint also shows better dispersion stability and small particle sizes, in parallel with hydrophobicity—enhancing the antimicrobial effects—which can be regulated by selection of a chain extender with a long hydrocarbon tail. In particular, the quaternizing of tertiary amines with different alkyl bromides was systematically investigated and indicated higher killing efficiency for bacterial and fungal strains with a longer alkyl chain [28]. The waterborne poly(methacrylate) suspensions were prepared with this type of antimicrobial hyper-branched emulsifier. Other cationic acrylate emulsions with antimicrobial copolymers were synthesized using alternative quaternary ammonium compounds or ammonium chloride derivates [29].

Chitin is a natural biopolymer recovered through extraction from crustaceans, and it has intrinsic antimicrobial properties. Chitosan has reactive amino groups on its pyranose ring and becomes a cationic polymer upon protonation of the amino groups. However, chitosan simply added to waterborne coatings cannot uniformly disperse. Furthermore, when chitosan-acid solution is added to waterborne coatings using acrylic emulsions, precipitates are formed because chitosan is a cationic polymer. Therefore, the incorporation of chitosan in paint formulations has been rarely considered, but one method for preparation of a hybridized chitin-acrylic emulsion was developed through emulsion polymerization and optimization of the pre-emulsification methods. When the products were applied as interior finishing coats, a reduction in formaldehyde content of the paint was observed [30]. The synthesis by emulsion polymerization of an acrylic resin with a quaternary ammonium salt (hybrid chitosan/acrylic resin emulsion) also resulted in better stability and less leaching due to better cross-linking between the acrylic resin particles [31]. Polyurethane paints could also be modified by the addition of chitosan/bentonite nanocomposites to improve antimicrobial properties [32]. As a more advanced approach, emulsion paints were modified through the addition of chitosan-grafted acrylic acid [33]. The chitosan-grafted acrylic acid was then used as an additive in a modified emulsion paint. The produce had higher density, more flexibility, better adhesion to the substrate and shorter drying times. The antimicrobial performance of the wet paint benefits from the presence of chitosan. It depends on the grafting efficiency and concentration of the grafted acrylic acid.

3.3. Bio-Engineered Enzymes and Peptide-Based Polymers

Enzymes that can be used as additives will assist in the inhibition of bacterial growth, e.g., through degradation of the cell wall (lysozymes), interaction with the biofilm or glycocalyx (alginate lyase) or generation of reactive oxygen species (glucose oxidase). Glucose oxidase favors the oxidation of glucose and the release of hydrogen peroxide, which creates damage to the cell wall and causes the destruction of lipids, proteins and sugars. The synergistic effects of bio-based biocides such as glucose oxidase and lysozymes added to traditional paints with MIT biocide were demonstrated [34]. They reduced the necessary concentrations of MIT.

Antimicrobial peptides (AMP) typically have a length of 6 or 7 amino acids and have inherent activity against certain bacteria, fungi and molds, depending on the peptide. Cationic peptides are designed with amino acids that have net positive charges (e.g., arginine, lysine and histidine) and hydrophobic amino acids (e.g., leucine and phenylalanine). The biocidal activity of AMP can be related to the permeation and disruption of the nuclear membrane. Screening of hexapeptides (AMP-6) and heptapeptides (AMP-7) as in-can preservatives for styrene-acrylic latex showed synergistic effects with concentrations of MIT below 15 ppm [35]: AMP-6 showed the best activity against fungi, whereas AMP-7 was satisfactory against both bacteria and fungi. A reduction in cellular metabolism of about 50% was observed when AMP-7 was added at 0.5–0.005 mg/mL. In future, it is believed that the peptide diversity in databases and access to large-scale synthetic production will allow researchers to adapt the antimicrobial activity of AMP for different paint systems.

3.4. Antimicrobial Nanocellulose

The addition of nanocellulose in waterborne coatings has been recently studied. Better service lives of the wood surfaces and higher mechanical resistance were found [36]. As a waterborne agent, it enables the control of paint rheology, in addition to improvements in antifouling and antibacterial properties after surface modification [37]. The antimicrobial properties of nanocellulose against bacteria, fungi and algae can be tuned with functional groups, such as aldehydes; quaternary ammonium; metal oxide nanoparticles; and chitosan [38]. Although mostly suggested for use in antimicrobial paint, nanocellulose is also compatible with waterborne acrylic dispersions, wherein the cellulose nanofibers (CNF) could be homogeneously mixed after facile mixing with small concentrations of aminopropyl-triethoxysilane under ultrasonic treatment and stirring [39]. The cationic or zwitterionic properties of modified nanocellulose after surface modifications with chemical grafting or adsorption of polyelectrolyte layers by electrostatic self-assembly may introduce antimicrobial properties. In contrast, cellulose nanocrystals (CNC) could be used more straightforwardly as a reinforcement filler in waterborne coatings. CNF would provide better mechanical reinforcement and barrier properties with their dense fibrillar network;

however, acrylate/CNF dispersions have high viscosity [40]. As such, nanocellulose could serve as a multifunctional bio-based ingredient for waterborne paint formulations.

4. Conclusions

In view of more strict regulations on biocides, the preservation of waterborne paint formulations is a huge challenge for coating and paint industries. Present solutions at the industrial scale focus on decreasing minimum inhibitory concentrations for isothiazolines through blending. Opportunities for disruptive and innovative technologies are offered by bio-inspired materials but will certainly need more development.

Author Contributions: Conceptualization, P.S. and P.C.; methodology, P.S.; formal analysis, P.S., J.B.; writing—original draft preparation, P.S.; writing—review and editing, P.S., J.B. and P.C.; project administration, J.B.; funding acquisition, P.C. All authors have read and agreed to the published version of the manuscript.

Funding: This research was funded by VLAIO, grant number HBC.2019.2493.

Institutional Review Board Statement: Not applicable.

Informed Consent Statement: Not applicable.

Data Availability Statement: Not applicable.

Conflicts of Interest: The authors declare no conflict of interest.

References

1. Lorenzen, J.; Poulsen, S. *Eco-Friendly Production of Waterborne Paint*; The Danish Environmental Protection Agency: Odense, Denmark, 2021.
2. Kharadi, N.; Mistry, R. *Economic Impact of Losing Effective In-Can Preservatives*; International Association for Soaps, Detergents and Maintenance Products: London, UK, 2018.
3. Silva, V.; Silva, C.; Soares, P.; Garrido, E.M.; Borges, F.; Garrido, J. Isothiazolinone biocides: Chemistry, biological and toxicity profiles. *Molecules* **2020**, *5*, 991. [CrossRef]
4. Müller, A.; Schmal, V.; Gschrei, S. *Survey on Alternatives for In-Can Preservation for Varnishes, Paints and Adhesives*; Federal Institute for Occupational Safety and Health: Berlin, Germany, 2020.
5. Paulus, W. Relationship between chemical structure and activity or mode of action of microbiocides. In *Directory of Microbiocides for the Protection of Materials*; Paulus, W., Ed.; Kluwer Academic Publishers: Dordrecht, The Netherlands, 2004; pp. 9–22.
6. Gillatt, J.; Julian, K.; Brett, K.; Goldbach, M.; Grohmann, J.; Heer, B.; Nichols, K.; Roden, K.; Rook, T.; Schubert, T.; et al. The microbial resistance of polymer dispersions and the efficacy of polymer dispersion biocides—A statistically validated method. *Int. Biodeterior. Biodegrad.* **2015**, *104*, 32–37. [CrossRef]
7. Wales, A.D.; Davies, R.H. Co-selection of resistance to antibiotics, biocides and heavy metals, and its relevance to foodborne pathogens. *Antibiotics* **2015**, *4*, 567–604. [CrossRef]
8. Kim, M.J.; Lim, K.B.; Lee, J.Y.; Kwack, S.J.; Kwon, Y.C.; Kang, J.S.; Kim, H.S.; Lee, B.M. Risk assessment of 5-chloro-2-methylisothiazol-3(2h)-one/2-methylisothiazol-3(2h)-one (cmit/mit) used as a preservative in cosmetics. *Toxicol. Res.* **2019**, *35*, 103–117. [CrossRef]
9. Betancur, J.; Browne, B.A. Innovating in-can preservatives depends on finding and testing the perfect blend. *Paint Coat. Ind.* **2021**, *17*, 8–15.
10. Rees, R. *Guidance on the Use of Globally-Relevant Modern Biocides*; Technical Papers; The Pressure Sensitive Tape Council: Lake Buena Vista, FL, USA, 2013; Volume 38, pp. 1–5.
11. BASF. *Industrial Product Preservation*; BASF: Ludwigshafen, Germany, 2000.
12. Brown, S.A. Past, present, and future options for preservative in coatings. *Coat. World* **2017**, *3*, 1–5.
13. Chervenak, M.C.; Konst, G.B.; Schwingel, W. Non-traditional use of the biocide DBNPA in coatings manufacture. *JCT Coat. Technol.* **2005**, *2*, 38–42.
14. Bellotti, N.; Romagnoli, R.; Quintero, C.; Dominguez-Wong, C.; Ruiz, F.; Deya, C. Nanoparticles as antifungal additives for indoor water borne paints. *Prog. Org. Coat.* **2015**, *86*, 33–40. [CrossRef]
15. Dankova, M.; Kalendova, A.; Machotova, J. Waterborne coatings based on acrylic latex containing nanostructured ZnO as an active additive. *J. Coat. Technol. Res.* **2020**, *17*, 517–529. [CrossRef]
16. Fiori, J.J.; Silva, L.L.; Picolli, K.C.; Ternus, R.; Ilha, J.; Decalton, F.; Mello, J.M.M.; Riella, H.G.; Fiori, M.A. Zinc oxide nanoparticles as antimicrobial additive for acrylic paint. *Mater. Sci. Forum* **2017**, *899*, 148–253. [CrossRef]
17. Kamal, H.B.; Antoniuos, M.S.; Mekewi, M.A.; Badawi, A.M.; Gabr, A.M.; El Bagdady, K. Nano ZnO/amine composites antimicrobial additives to acrylic paints. *Egypt J. Petrol.* **2015**, *24*, 397–404. [CrossRef]

18. Steinerova, D.; Kalendova, A.; Machotova, J.; Pejchalova, M. Environmentally friendly water-based self-crosslinking acrylate dispersion containing magnesium nanoparticles and their films exhibiting antimicrobial properties. *Coatings* **2020**, *10*, 340. [CrossRef]
19. Dileep, P.; Jacob, S.; Narayanankutty, S.K. Functionalized nanosilica as an antimicrobial additive for waterborne paints. *Prog. Org. Coat.* **2020**, *142*, 105574. [CrossRef]
20. Hendessi, S.; Sevinins, E.B.; Unal, S.; Cebeci, F.C.; Menceloglu, Y.Z.; Unal, H. Antibacterial sustained-release coatings from halloysite nanotubes/waterborne polyurethanes. *Prog. Org. Coat.* **2015**, *101*, 253–261. [CrossRef]
21. Stanojevick-Nikolic, S.; Dimic, G.; Mojovic, L.; Pejin, J.; Djukic-Vukovic, A.; Kocic-Tanackov, S. Antimicrobial activity of lactic acid against pathogen and spoilage microorganisms. *J. Food Proc. Pres.* **2015**, *40*, 990–998. [CrossRef]
22. Pasricha, A.; Bhalla, P.; Sharma, K.B. Evaluation of lactic acid as an antibacterial agent. *Indian J. Dermatol. Venereol. Leprol.* **1979**, *45*, 159–161. [PubMed]
23. Amrouche, T.; Noll, K.S.; Wang, Y.; Huang, Q.; Chikindas, M.L. Antibacterial activity of subtilosin alone and combined with curcumin, poly-lysine and zinc lactate against listeria monocytogenes strains. *Probiotics Antimicrob. Prot.* **2010**, *2*, 250–257. [CrossRef]
24. Gómez-García, M.; Sol, C.; de Nova, P.J.; Puyalto, M.; Mesas, L.; Puente, H.; Mencía-Ares, Ó.; Miranda, R.; Argüello, H.; Rubio, P.; et al. Antimicrobial activity of a selection of organic acids, their salts and essential oils against swine enteropathogenic bacteria. *Porc. Heatlh Manag.* **2019**, *5*, 32. [CrossRef]
25. Ribeiro, A.M.; Carrasco, L.D. Cationic antimicrobial polymers and their assemblies. *Int. J. Mol. Sci.* **2013**, *14*, 9905–9946.
26. Kamaruzzaman, N.F.; Tan, L.P.; Hamdan, R.H.; Choong, S.S.; Woing, W.K.; Gibson, A.J.; Chivu, A.; Pina, M. Antimicrobial polymers: The potential replacement of existing antibiotics. *Int. J. Mol. Sci.* **2019**, *20*, 2747. [CrossRef]
27. Wang, Y.; Chen, R.; Li, T.; Ma, P.; Zhang, H.; Du, M.; Chen, M.; Dong, W. Antimicrobial waterborne polyurethanes based on quaternary ammonium compounds. *Ind. Eng. Chem. Res.* **2020**, *59*, 458–463. [CrossRef]
28. Zhao, P.; Mecozzi, F.; Wessel, S.; Fieten, B.; Driesse, M.; Woudstra, W.; Busscher, H.J.; Mei, H.C.; Loontjens, T.J.A. Preparation and evaluation of antimicrobial hyperbranched emulsifiers for waterborne coatings. *Langmuir* **2019**, *35*, 5779–5786. [CrossRef] [PubMed]
29. Wu, Y.; Gan, J.; Yang, F.; Zhang, H.; Wang, W. Preparation and antibacterial properties of waterborne UV cured coating modified by quaternary ammonium compounds. *J. Appl. Polym. Sci.* **2021**, *138*, 50426. [CrossRef]
30. Wada, T.; Uragami, T.; Matoba, Y. Chitosan-hybridized acrylic resins prepared in emulsion polymerizations and their application as interior finishing coatings. *JCT Res.* **2005**, *2*, 577–592. [CrossRef]
31. Wada, T.; Yasuda, M.; Yako, H.; Matoba, Y.; Uragami, T. Preparation and characterization of hybrid quaternized chitosan/acrylic resin emulsions and their films. *Macromol. Mater. Eng.* **2007**, *292*, 147–154. [CrossRef]
32. Rihayat, T.; Satriananda, S.; Nurhanifa, R. Influence of coating polyurethane with mixture of bentonite and chitosan nanocomposites. *AIP Conf. Proc.* **2018**, *2049*, 020020.
33. Abolude, O.I. Modification of Emulsion Paint Using Chitosan-Grafted Acrylic Acid. Master's Thesis, Ahmadu Bello University, Zaria, Nigeria, 2016.
34. Hodges, T.W.; Kemp, L.K.; McInnes, B.M.; Wilhelm, K.L.; Hurt, J.D.; McDaniel, S.; Rawlins, J.W. Proteins and Peptides as Replacements for Traditional Organic Preservatives. *Coat. Technol.* **2018**, *15*, 45–50.
35. McDaniel, S.; McInnis, B.M.; Hurt, J.D.; Kemp, L.K. Biotechnology meets coatings preservation. *Coat. World* **2019**, *12*, 33–42.
36. Kluge, M.; Veigel, S.; Pinkl, S.; Henniges, U.; Zollfrank, C.; Rössler, A.; Gindl-Altmutter, W. Nanocellulosic fillers for waterborne wood coatings: Reinforcement effect on free-standing coating films. *Wood Sci. Technol.* **2017**, *51*, 601–613. [CrossRef]
37. Aguilar-Sanchez, A.; Jalvo, B.; Mautner, A.; Nameer, S.; Pöhler, T.; Tammelin, T.; Mathew, A.J. aterborne nanocellulose coatings for improving the antifouling and antibacterial proper-ties of polyethersulfone membranes. *J. Membrane Sci.* **2021**, *620*, 118842. [CrossRef]
38. Norrahim, M.; Nurazzi, N.M.; Jenol, M.A.; Farid, M.A.; Janudin, N.; Ujang, F.A.; Yasim-Anuar, T.A.; Najmuddine, S.U.; Ilyasf, R.A. Emerging development of nanocellulose as an antimicrobial material: An overview. *Mater. Adv.* **2021**, *2*, 3538–3551. [CrossRef]
39. Tan, Y.; Liu, Y.; Chen, W.; Liu, Y.; Wang, Q.; Li, J.; Yu, H. Homogeneous dispersion of cellulose nanofibers in waterborne acrylic coatings with improved properties and unreduced transparency. *ACS Sustain. Chem. Eng.* **2016**, *4*, 3766–3772. [CrossRef]
40. Hassan, M.L.; Fadel, S.M.; Hassan, E.A. Acrylate/nanofibrillated cellulose nanocomposites and their use for paper coating. *J. Nanomater.* **2018**, *2018*, 4963834. [CrossRef]

Abstract

Biopolymer-Based Hydrogels for 3D Bioprinting [†]

Ahmed Fatimi [1,2,*], Oseweuba Valentine Okoro [3] and Amin Shavandi [3]

[1] Department of Chemistry, Polydisciplinary Faculty of Beni-Mellal (FPBM), Sultan Moulay Slimane University (USMS), P.O. Box 592 Mghila, Beni-Mellal 23000, Morocco

[2] ERSIC, Polydisciplinary Faculty of Beni-Mellal (FPBM), Sultan Moulay Slimane University(USMS), P.O. Box 592 Mghila, Beni-Mellal 23000, Morocco

[3] BioMatter Unit, École Polytechnique de Bruxelles, Université Libre de Bruxelles (ULB), 1050 Brussels, Belgium; oseweubaokoro@gmail.com (O.V.O.); amin.shavandi@ulb.be (A.S.)

[*] Correspondence: a.fatimi@usms.ma

[†] Presented at the 2nd International Online Conference on Polymers Science–Polymers and Nanotechnology for Industry 4.0, 1–15 November 2021; Available online: https://iocps2021.sciforum.net/.

Citation: Fatimi, A.; Okoro, O.V.; Shavandi, A. Biopolymer-Based Hydrogels for 3D Bioprinting. *Mater. Proc.* **2021**, *7*, 19. https://doi.org/10.3390/IOCPS2021-11284

Academic Editor: Carlo Santulli

Published: 1 November 2021

Publisher's Note: MDPI stays neutral with regard to jurisdictional claims in published maps and institutional affiliations.

Abstract: Three-dimensional (3D) bioprinting is an emerging technology that could be used in the generation of 3D cellular structures for tissue engineering applications. The interest in this technology is due to its capacity to enable the fabrication of precise 3D constructs composed of biomaterials laden with living cells, biomolecules, and nutrients. The process involving the deposition of cell-laden biomaterials or bioinks on a substrate is referred to as bioprinting. This bioprinting process can be used in the fabrication of living tissues and functional organs suitable for transplantation into the human body. Notably, the viability of utilising a bioink for bioprinting is dependent on its functionality, mechanical properties, printability, and biocompatibility. The bioink must also be able to provide cells with a stable environment for attachment, proliferation, and differentiation. To promote the sufficiency of bioinks in 3D bioprinting, several researchers have investigated pathways to enhance ink properties to meet bioprinting requirements, with several synthetic and natural hydrogels developed. These hydrogels are matrices made up of a network of hydrophilic polymers that absorb biological fluids. They can be created from a large number of water-soluble biopolymers including proteins and polysaccharides. The 3D structure of these hydrogels is due to the presence of structural crosslinks that are maintained the environmental fluid. The elasticity of these structures and the presence of a large amount of water enable the hydrogel to adequately mimic biological tissues. Recognising the importance of hydrogels in 3D bioprinting and its potential wide range of tissue engineering applications, the current study therefore investigated major physicochemical parameters that may affect the printability and biocompatibility of biopolymer-based hydrogels. Approaches employed in maintaining structural integrity of the hydrogel, via the application of crosslinking methods, were comprehensively discussed with explorations of the status of the formulation and the use of biopolymer-based hydrogels for 3D bioprinting is also presented.

Keywords: 3D bioprinting; hydrogels; biopolymers; bioinks

Supplementary Materials: The following supporting information can be downloaded at: https://www.mdpi.com/article/10.3390/IOCPS2021-11284/s1.

Author Contributions: Conceptualization, A.F. and A.S.; methodology, A.F., A.S. and O.V.O.; validation, A.F., A.S. and O.V.O.; investigation, A.F., A.S. and O.V.O.; writing—original draft preparation, A.F., A.S. and O.V.O.; writing—review and editing, A.F., A.S. and O.V.O. All authors have read and agreed to the published version of the manuscript.

Funding: This research received no external funding.

Institutional Review Board Statement: Not applicable.

Informed Consent Statement: Not applicable.

Data Availability Statement: Not applicable.

Conflicts of Interest: The authors declare no conflict of interest.

Proceeding Paper

Preparation and Characterisation of PBAT-Based Biocomposite Materials Reinforced by Protein Complex Microparticles [†]

Elena Togliatti [1,2,*], Cosimo C. Laporta [1], Maria Grimaldi [3], Olimpia Pitirollo [3], Antonella Cavazza [3], Diego Pugliese [2,4], Daniel Milanese [1,2] and Corrado Sciancalepore [1,2]

[1] DIA, Dipartimento di Ingegneria e Architettura, Università di Parma, Parco Area delle Scienze 181/A, 43124 Parma, Italy; cosimocataldo.laporta@studenti.unipr.it (C.C.L.); daniel.milanese@unipr.it (D.M.); corrado.sciancalepore@unipr.it (C.S.)

[2] INSTM, Consorzio Interuniversitario Nazionale di Scienza e Tecnologia dei Materiali, Via G. Giusti 9, 50121 Firenze, Italy; diego.pugliese@polito.it

[3] SCVSA, Dipartimento di Scienze Chimiche, Della Vita e della Sostenibilità Ambientale, Università di Parma, Parco Area delle Scienze 17/A, 43124 Parma, Italy; daianagrimaldi@hotmail.it (M.G.); olimpia.pitirollo@unipr.it (O.P.); antonella.cavazza@unipr.it (A.C.)

[4] DISAT, Dipartimento di Scienza Applicata e Tecnologia, Politecnico di Torino, Corso Duca degli Abruzzi 24, 10129 Torino, Italy

* Correspondence: elena.togliatti@studenti.unipr.it

[†] Presented at the 2nd International Online Conference on Polymer Science—Polymers and Nanotechnology for Industry 4.0, 1–15 November 2021; Available online: https://iocps2021.sciforum.net/.

Abstract: In this work, new biodegradable composite materials based on poly (butylene adipate terephthalate) (PBAT) reinforced with zein–TiO$_2$ complex microparticles were prepared and characterised by electron microscopy and tensile and dynamic-mechanical tests. The composite pellets were prepared by solvent casting with different filler contents, namely 0, 5.3, 11.1 and 25 part per hundred resin (phr), to modify and modulate the properties of the final materials. Scanning electron microscopy (SEM) images showed homogeneous dispersion of the filler, without microparticles aggregation or phase separation between filler and matrix, suggesting a good interphase adhesion. According to tensile tests, Young's modulus showed an improvement in the rigidity and the yield stress presented an increasing trend, with opposite behaviour compared to other composites. Dynamic-mechanical analysis (DMA) results exhibited increasing storage modulus values, confirming a greater rigidity with a higher filler percentage. The glass transition temperature showed a slightly increasing trend, meaning the presence of an interaction between the two phases of the composite materials. Overall, the produced PBAT composites showed similar properties to low-density polyethylene (LDPE), proving to be promising and more sustainable alternatives to traditional polymers commonly adopted in agri-food fields.

Keywords: biopolymers; biocomposites; poly (butylene adipate terephthalate); protein complex; characterisation

Citation: Togliatti, E.; Laporta, C.C.; Grimaldi, M.; Pitirollo, O.; Cavazza, A.; Pugliese, D.; Milanese, D.; Sciancalepore, C. Preparation and Characterisation of PBAT-Based Biocomposite Materials Reinforced by Protein Complex Microparticles. *Mater. Proc.* **2021**, *7*, 20. https://doi.org/10.3390/IOCPS2021-12019

Academic Editor: Ana María Díez-Pascual

Published: 2 December 2021

Publisher's Note: MDPI stays neutral with regard to jurisdictional claims in published maps and institutional affiliations.

1. Introduction

Among materials for packaging, plastics are the most widely used, thanks to their lightness, good mechanical behaviour, barrier properties and low cost, among others [1]. Amongst traditional plastics, the most employed are polypropylene (PP), high-density polyethylene (HDPE), low-density polyethylene (LDPE), polyethylene terephthalate (PET) and polystyrene (PS), which however are not eco-sustainable due to the problems related to their end-of-life disposal [2].

In the last few decades, increasing attention has been devoted to the study and employment of bioplastics in order to reduce the environmental impact and increase sustainability. Since bioplastics generally present poorer properties when compared to traditional plastics, the realisation of composite materials represents a valid way to improve and modulate their characteristics. The major downside of biopolymers being their high cost [3], the use of natural biodegradable fillers is a possible solution to reduce the production costs and at the same time to preserve their degradability [4].

Poly (butylene adipate terephthalate) (PBAT) is a 100% biodegradable polymer produced from fossil resources [5], although recently it has been reported that its monomers can be obtained from renewable sources [6–8]. PBAT presents similar properties and processability to polyethylene (PE), especially high flexibility [9,10].

Zein is a prolammine protein which can be extracted in pure form from corn. Its use in polymers has been studied since the 20th century [11] as it is considered a safe biocompatible and biodegradable material [12]. Zein can be formed into films and displays good barrier properties, thanks to its hydrophobic nature [13]. However, protein films are usually tough and brittle and cannot be used as is, but protein in the form of particles can be used as the reinforcing phase in the realisation of composites which are based on a flexible polymer matrix [14], such as PBAT in this case.

Of late, zein has been functionalised in protein–TiO_2 complexes for packaging, environmental and medical applications [15].

The aim of this work was to design and fabricate biocomposites based on PBAT loaded with microparticles of a zein–TiO_2 complex. The so-obtained composites have been characterised in terms of their mechanical and dynamic-mechanical properties.

2. Materials and Methods

PBAT (MAgMa Spa) pellets were dissolved into pure chloroform. The zein–TiO_2 complex had been previously prepared with a composition of 50–50 wt%, by first dissolving the zein (Sigma-Aldrich, St. Louis, MO, USA) in ethanol at 50 °C and then adding TiO_2 (Carlo Erba, Emmendingen, Germany) under constant stirring until a homogeneous phase was obtained. After casting and ethanol evaporation, the recovered material was milled and sieved at 25 μm, and the so-obtained powder was homogeneously dispersed into the polymer solution at the concentrations of 0, 5.3, 11.1 and 25 phr. After solvent evaporation, the obtained films were used for the production of different loaded composite samples, named PBAT, PBAT + 5.3P, PBAT + 11.1P and PBAT + 25P, respectively.

Dumbbell specimens, model 1BA according to the UNI EN ISO 527 standard, of each composite were produced by injection moulding and their mechanical properties were characterised.

Uniaxial tensile test (UTT) results allowed the evaluation of the characteristic parameters, such as Young's modulus E, yield stress σ_Y, elongation at break ε_B, stress at break σ_B and toughness T.

DMA measurements were carried out according to ASTM D7028 standard with a single cantilever clamp for the determination of the storage modulus (E'), the loss modulus (E'') and the loss factor ($\tan\delta$).

SEM images were acquired to investigate the internal microstructure of the composites, by means of a field emission gun SEM (FESEM, Nova Nano SEM 450, FEI company, Hillsboro, OR, USA). SEM was performed on the central cross-section of the specimens obtained through a cryo-fracture.

3. Results and Discussion

In Figure 1, SEM images at different magnifications of PBAT and PBAT + 25P are reported as representative samples. The morphology of the filler particles (as visible in detail in Figure 1d) emphasises the protein–TiO_2 complex nature, showing bright-white areas corresponding to the TiO_2 portion and a greyer part representing the zein protein.

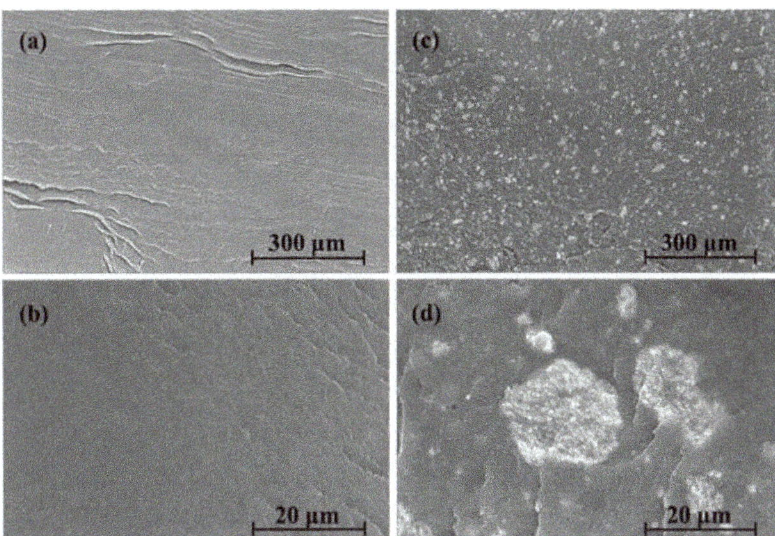

Figure 1. SEM images of PBAT (**a,b**) and PBAT + 25P (**c,d**) at different magnifications, as representative samples.

The images display homogeneous dispersion of the protein complex within the polymer matrix, even at a high concentration of the filler, with no aggregation of the particles and no phase separation (Figure 1b). Indeed, a region with an intermediate shade of grey is visible at the grain boundaries. This suggests the formation of an actual interfacial layer bonding the PBAT matrix to the protein due to the presence in the protein structure of both polar and non-polar functional groups [16], able to interact with the polymer macromolecules. Therefore, good adhesion and interaction between the phases can be supposed.

UTT results on the prepared biocomposites indicated a pronounced increase, up to 47%, in the E modulus with increasing filler content. An important result is the increasing trend shown by σ_Y, significantly different from what is traditionally displayed by other composite materials [17]. The obtained results can be interpreted as additional evidence of the good interaction between the phases involved in the biocomposite [18]. The characteristic parameters ε_B, σ_B and T showed a decrease with increasing filler content, related to the increased stiffening of the composites. The representative values are reported in Table 1.

Table 1. Young's modulus (E), yield stress (σ_Y), stress at break (σ_B), elongation at break (ε_B) and toughness (T) values of poly (butylene adipate terephthalate) (PBAT) and protein complex composites.

Sample	E (MPa)	σ_Y (MPa)	σ_B (MPa)	ε_B (MPa)	T (MJ/m^3)
PBAT	126 ± 12	8.1 ± 0.2	13 ± 1	4.0 ± 0.8	45 ± 6
PBAT + 5.3P	131 ± 10	8.4 ± 0.2	12 ± 1	3.7 ± 0.3	35 ± 4
PBAT + 11.1P	149 ± 4	8.8 ± 0.2	11 ± 1	3.5 ± 0.2	32 ± 3
PBAT + 25P	186 ± 11	8.9 ± 0.1	9 ± 1	2.7 ± 0.2	22 ± 2

The E' modulus obtained from DMA analysis as a function of the temperature exhibited a linear increase with increasing filler content in the composite, thus confirming the stiffening effect obtained by the addition of high amounts of filler (Figure 1c).

Other authors have investigated similar biocomposite systems, based on biopolymers reinforced with natural filler particles at different concentrations [19], finding a similar increasing behaviour of E' compared to the system studied in the present work. The enhanced modulus in composite materials can be attributed to the restricted mobility of the polymer chains due to the physical presence of the filler particles and to the chemical interaction at the interface between the polymer and the particles [20].

The glass transition temperature (T_g) was calculated as the temperature corresponding to the peak of the tanδ curves, defined by the ratio between E'' and E' moduli. The values of T_g for the different composites show a slight increase as the filler content increases, confirming the interaction between matrix and filler, as observed in other composite systems [21]. Table 2 displays the discussed results of DMA tests.

Table 2. DMA representative results of PBAT and protein complex composites.

Sample	E' @ 0 °C (MPa)	E' @ 20 °C (MPa)	E' @ 40 °C (MPa)	T_g (°C)
PBAT	273 ± 60	202 ± 50	155 ± 50	-20.4 ± 0.9
PBAT + 5.3P	297 ± 60	224 ± 70	182 ± 60	-20.3 ± 0.8
PBAT + 11P	319 ± 70	239 ± 60	192 ± 70	-18.3 ± 0.5
PBAT + 25P	395 ± 50	294 ± 50	235 ± 60	-17.9 ± 0.5

4. Conclusions

Among the different biopolymers, PBAT is one of the most studied and promising biodegradable plastic materials. In this work, it was employed in the fabrication of biocomposites reinforced with a zein–TiO$_2$ complex at different concentration. The addition of different amounts of filler enabled modulation of the material properties.

The filler particles were homogeneously dispersed, as emerged from SEM images of the analysed samples, and with the presence of an interface connecting layer between the protein complex and the polymer matrix. The protein complex appeared to have a stiffening effect on the polymer matrix, with an increase of the E and σ_Y, suggesting, therefore, an effective good interfacial interaction between the phases.

The stiffening effect was confirmed by the increasing trend observed in the E' modulus calculated from DMA analysis. Moreover, T_g values increased with increasing filler content, validating the hypothesis of an interface layer bonding the matrix and the reinforcing particles.

According to the obtained results, the biocomposites can be considered as a valid and more sustainable alternative to the non-biodegradable, fossil-based plastics generally used in the packaging field, such as LDPE.

Author Contributions: Conceptualisation, C.S.; methodology, C.S.; formal analysis, C.C.L.; investigation, E.T., C.C.L., M.G., O.P. and D.P.; resources, A.C.; data curation, C.C.L.; writing—original draft preparation, E.T.; writing—review and editing, D.M. and C.S.; supervision, D.M. and C.S. All authors have read and agreed to the published version of the manuscript.

Funding: This research received no external funding.

Data Availability Statement: Not applicable.

Conflicts of Interest: The authors declare no conflict of interest.

References

1. Piergiovanni, L.; Limbo, S. Food packaging. In *Materiali, Tecnologie e Qualità Degli Alimenti*, 1st ed.; Springer: Berlin, Germany, 2010; pp. 1–576.
2. Andrady, A.L.; Neal, M.A. Applications and societal benefits of plastics. *Philos. Trans. R. Soc. London, Ser. B* **2009**, *364*, 1977–1984. [CrossRef] [PubMed]
3. Harrison, J.P.; Boardman, C.; O'Callaghan, K.; Delort, A.-M.; Song, J. Biodegradability standards for carrier bags and plastic films in aquatic environments: A critical review. *R. Soc. Open Sci.* **2018**, *5*, 171792. [CrossRef]

4. Xiong, S.-J.; Pang, B.; Zhou, S.-J.; Li, M.-K.; Yang, S.; Wang, Y.-Y.; Shi, Q.; Wang, S.-F.; Yuan, T.-Q.; Sun, R.-C. Economically competitive biodegradable PBAT/lignin composites: Effect of lignin methylation and compatibilizer. *ACS Sustain. Chem. Eng.* **2020**, *8*, 5338–5346. [CrossRef]

5. Ferreira, F.V.; Cividanes, L.S.; Gouveia, R.F.; Lona, L.M.F. An overview on properties and applications of poly(butyleneadipate-co-terephthalate)–PBAT based composites. *Polym. Eng. Sci.* **2019**, *59*, E7–E15. [CrossRef]

6. Volanti, M.; Cespi, D.; Passarini, F.; Neri, E.; Cavani, F.; Mizsey, P.; Fozer, D. Terephthalic acid from renewable sources: Early-stage sustainability analysis of a bio-PET precursor. *Green Chem.* **2019**, *21*, 885–896. [CrossRef]

7. Skoog, E.; Shin, J.H.; Saez-Jimenez, V.; Mapelli, V.; Olsson, L. Biobased adipic acid—The challenge of developing the production host. *Biotechnol. Adv.* **2018**, *36*, 2248–2263. [CrossRef] [PubMed]

8. Silva, R.G.C.; Ferreira, T.F.; Borges, É.R. Identification of potential technologies for 1,4-Butanediol production using prospecting methodology. *J. Chem. Technol. Biotechnol.* **2020**, *95*, 3057–3070. [CrossRef]

9. Kijchavengkul, T.; Auras, R.; Rubino, M.; Ngouajio, M.; Fernandez, R.T. Assessment of aliphatic–aromatic copolyester biodegradable mulch films. *Part I: Field study. Chemosphere* **2008**, *71*, 942–953. [PubMed]

10. Jian, J.; Xiangbin, Z.; Xianbo, H. An overview on synthesis, properties and applications of poly(butylene-adipate-co-terephthalate)–PBAT. *Adv. Ind. Eng. Polym. Res.* **2020**, *3*, 19–26. [CrossRef]

11. Lawton, J.W. Zein: A history of processing and use. *Cereal Chem.* **2002**, *79*, 1–18. [CrossRef]

12. Oymaci, P.; Altinkaya, S.A. Improvement of barrier and mechanical properties of whey protein isolate based food packaging films by incorporation of zein nanoparticles as a novel bionanocomposite. *Food Hydrocolloids* **2016**, *54*, 1–9. [CrossRef]

13. Oh, J.-H.; Wang, B.; Field, P.D.; Aglan, H.A. Characteristics of edible films made from dairy proteins and zein hydrolysate cross-linked with transglutaminase. *Int. J. Food Sci. Technol.* **2004**, *39*, 287–294. [CrossRef]

14. Kadam, D.M.; Thunga, M.; Srinivasan, G.; Wang, S.; Kessler, M.R.; Grewell, D.; Yu, C.; Lamsal, B. Effect of TiO2 nanoparticles on thermo-mechanical properties of cast zein protein films. *Food Packag. Shelf Life* **2017**, *13*, 35–43. [CrossRef]

15. Anaya-Esparza, L.M.; Villagrán-de la Mora, Z.; Rodríguez-Barajas, N.; Sandoval-Contreras, T.; Nuño, K.; López-de la Mora, D.A.; Pérez-Larios, A.; Montalvo-González, E. Protein–TiO2: A functional hybrid composite with diversified applications. *Coatings* **2020**, *10*, 1194. [CrossRef]

16. Guo, X.; Ren, C.; Zhang, Y.; Cui, H.; Shi, C. Stability of zein-based films and their mechanism of change during storage at different temperatures and relative humidity. *J. Food Process. Preserv.* **2020**, *44*, e14671. [CrossRef]

17. Százdi, L.; Pukánszky, B.; Vancso, G.J.; Pukánszky, B. Quantitative estimation of the reinforcing effect of layered silicates in PP nanocomposites. *Polymer* **2006**, *47*, 4638–4648. [CrossRef]

18. Pegoretti, A.; Dorigato, A.; Penati, A. Tensile mechanical response of polyethylene—Clay nanocomposites. *eXPRESS Polym. Lett.* **2007**, *1*, 123–131. [CrossRef]

19. Nanni, A.; Messori, M. Thermo-mechanical properties and creep modelling of wine lees filled Polyamide 11 (PA11) and Polybutylene succinate (PBS) bio-composites. *Compos. Sci. Technol.* **2020**, *188*, 107974. [CrossRef]

20. Dorigato, A.; D'Amato, M.; Pegoretti, A. Thermo-mechanical properties of high density polyethylene—Fumed silica nanocomposites: Effect of filler surface area and treatment. *J. Polym. Res.* **2012**, *19*, 9889. [CrossRef]

21. Kang, S.; Hong, S.I.; Choe, C.R.; Park, M.; Rim, S.; Kim, J. Preparation and characterization of epoxy composites filled with functionalized nanosilica particles obtained via sol–gel process. *Polymer* **2001**, *42*, 879–887. [CrossRef]

materials proceedings

Abstract

A Novel Treatment Tool for PLA-Based Encapsulation Systems [†]

Konstantina Chronaki [1], Angeliki Mytara [1], Constantine D. Papaspyrides [1], Konstantinos Beltsios [2] and Stamatina Vouyiouka [1,*]

[1] Laboratory of Polymer Technology, School of Chemical Engineering, National Technical University of Athens, Zographou Campus, 157 80 Athens, Greece; kchronaki@mail.ntua.gr (K.C.); mytara.a@live.com (A.M.); papaspcd@gmail.com (C.D.P.)

[2] School of Chemical Engineering, National Technical University of Athens, Zographou Campus, 157 80 Athens, Greece; kgbelt@mail.ntua.gr

* Correspondence: mvuyiuka@central.ntua.gr

[†] Presented at the 2nd International Online Conference on Polymer Science-Polymers and Nanotechnology for Industry 4.0, 1–15 November 2021; Available online: https://iocps2021.sciforum.net/.

Abstract: Active-compound encapsulation in polymeric carriers is a widely used technology as it protects and improves the physical characteristics of the active compound and controls its delivery. The effectiveness of polymeric microcapsules (MCs) depends on the barrier properties of the polymeric shell; for a given polymer, the latter properties are affected by its molecular weight (MW) and crystallinity (x_c). The aim of this study was to modify the MW and x_c of the MCs shell via solid state polymerization (SSP). SSP may take place in the amorphous regions of the polymer upon heating at temperatures higher than the glass transition point (T_g) but lower than the onset of melting (T_m). Poly(lactic acid) (PLA) was chosen as the polymeric carrier and coumarin 6 as the encapsulated compound. PLA is a biobased and biodegradable polymer that is widely used in drug delivery systems, and coumarin 6 is a fluorescent hydrophobic drug that can be used as model compound. The effectiveness of SSP as a post-encapsulation tool was proven for blank PLA MCs of two molecular weights (MW = 50,000 g mol^{-1} and 20,000 g mol^{-1}). SSP led to a 40–50% enhancement of the weight-average molecular weight of the polymeric shell and to an enhancement, from 40% to up to 70%, of the mass fraction crystallinity in the case of the low-MW MCs. In an attempt to transfer the gained knowledge to the encapsulation systems, coumarin-6-loaded MCs were prepared. The average size of the MCs was measured at 502 nm with a polydispersity index of 1.6 while the encapsulation efficiency was found to be 15% for a drug loading of 10%. UV–Vis measurements showed that the compound was fully released after 10 days. Coumarin 6 was found to be thermally stable at temperatures used for SSP, while the study of SSP application in the case of loaded MCs is in progress.

Keywords: encapsulation systems; emulsification-solvent evaporation; microcapsules; solid-state polymerization; poly(lactic acid); coumarin 6

Citation: Chronaki, K.; Mytara, A.; Papaspyrides, C.D.; Beltsios, K.; Vouyiouka, S. A Novel Treatment Tool for PLA-Based Encapsulation Systems. *Mater. Proc.* **2021**, *7*, 21. https://doi.org/10.3390/IOCPS2021-11268

Academic Editor: Ana María Díez-Pascual

Published: 1 November 2021

Supplementary Materials: The following supporting information can be downloaded at: https://www.mdpi.com/article/10.3390/IOCPS2021-11268/s1.

Funding: This research was co-financed by Greece and the European Union (European Social Fund-ESF) through the Operational Programme «Human Resources Development, Education and Lifelong Learning 2014–2020» in the context of the project "Encapsulation of active compounds in polymeric microcapsules: Modifying properties via Solid State Polymerization" (MIS 5049531).

Proceeding Paper

Assessment of the Optical Properties of a Graphene–Poly(3-hexylthiophene) Nanocomposite Applied to Organic Solar Cells [†]

Lara Velasco Davoise [1],*, Rafael Peña Capilla [2] and Ana M. Díez-Pascual [1]

[1] Universidad de Alcalá, Facultad de Ciencias, Departamento de Química Analítica, Química Física e, Ingeniería Química, Ctra. Madrid-Barcelona, Km. 33.6, 28805 Alcalá de Henares, Madrid, Spain; am.diez@uah.es

[2] Universidad de Alcalá, Departamento de Teoría de la Señal y Comunicaciones, Ctra. Madrid-Barcelona, Km. 33.6, 28805 Alcalá de Henares, Madrid, Spain; rafael.pena@uah.es

* Correspondence: lara.velasco@uah.es

[†] Presented at the 2nd International Online Conference on Polymers Science—Polymers and Nanotechnology for Industry 4.0, 1–15 November 2021; Available online: https://iocps2021.sciforum.net/.

Abstract: Poly(3-hexylthiophene) (P3HT) is a p-type organic semiconductor and is intrinsically a donor material. It is one of the most attractive polymers because of its high electrical conductivity and solubility in various solvents. However, its carrier mobility is considered low when compared to that of inorganic semiconductors. In this work, it will be shown how the addition of different graphene (G) content tailors the principal optical and electronical parameters of P3HT, such as the conductivity, the bandgap, the hole collection properties, the carrier mobility, the refractive index, and the extinction coefficient. In particular, the conductivity, the hole collection properties and the carrier mobility are enhanced, and the bandgap is reduced with increasing graphene content.

Keywords: nanocomposite; refractive index; extinction coefficient; graphene

Citation: Davoise, L.V.; Capilla, R.P.; Díez-Pascual, A.M. Assessment of the Optical Properties of a Graphene–Poly(3-hexylthiophene) Nanocomposite Applied to Organic Solar Cells. *Mater. Proc.* **2021**, *7*, 22. https://doi.org/10.3390/IOCPS2021-11241

Academic Editor: Marina Patricia Arrieta Dillon

Published: 30 October 2021

Publisher's Note: MDPI stays neutral with regard to jurisdictional claims in published maps and institutional affiliations.

1. Introduction

New photovoltaic energy technologies can contribute to environmentally friendly, renewable energy production and a reduction in the carbon dioxide emission associated with fossil fuels and biomass. A new photovoltaic technology, organic solar cells, is based on conjugated polymers and molecules. Organic solar cells have attracted considerable attention in the past few years, owing to their potential for providing environmentally safe, flexible, lightweight, and inexpensive devices [1]. In this context, efforts are mostly concentrated on increasing power conversion efficiencies and reducing the costs of materials and processing conditions.

Poly(3-hexylthiophene) (P3HT) is a p-type semiconductor and is intrinsically a donor material. P3HT is a regioregular polymer, which means that each repeating unit is derived from the same isomer of the monomer, and is one of the most attractive polymers available because of its high electrical conductivity and solubility in various solvents. Moreover, P3HT has a high absorption coefficient (in the order of 10^5 cm^{-1}) and can absorb more than 95% of the solar spectrum over a wavelength range of 450–600 nm when deposited in a 240 nm thick film. This makes it very attractive for organic optoelectronic devices [2]. Moreover, P3HT possesses several other advantages such as solution processability and easy and low-cost fabrication. It also possesses one of the highest electrical mobilities of the known conjugated polymers: 0.2 cm^2/(V·s). However, this mobility is considered low when compared to those of inorganic semiconductors (Figure 1) [3,4].

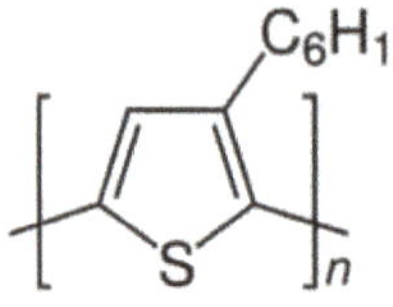

$$C_6H_{13}$$

Figure 1. Scheme of P3HT monomeric unit [4].

Novel acceptor materials such as graphene (G)—a one-atom-thick sheet of sp^2 hybridized carbon atoms arranged in a hexagonal lattice, first discovered in 2004 by Novoselov et al. [5]—have been intensively investigated given their outstanding electrical, optical, chemical, and mechanical properties. Graphene has an extraordinary electrical mobility of 15000 $cm^2/(V \cdot s)$, a high specific surface area (ca. 2600 m^2/g), very high electrical conductivity (up to 6000 S/cm), and a transparency of more than 70% over the spectral range of 1000–3000 nm [3]. A single sheet of graphene has superior mechanical properties because of the strong p-bond in its honeycomb crystal lattice structure, with a tensile strength of 130 GPa and a Young's modulus close to 1 TPa [3,6].

However, the very low responsivity due to its weak light absorption and fast recombination rate has limited the sensitivity of graphene light-sensing devices [3]. Significant efforts have been applied to increase its absorption, among which a feasible way is the combination with light-absorbing materials [6].

Several methods for the preparation of P3HT/G nanocomposites have been reported, including solution mixing and in situ polymerization [7,8]. In situ polymerization consists of mixing nanofillers with a liquid monomer or a precursor of a low molecular weight. When a homogenous mixture is attained, polymerization is initiated by the addition of an appropriate initiator, which is exposed to a source of heat, radiation, etc. [9]. Polymerization is carried out by adjusting the temperature and time. It is a very effective method that allows carbon-based nanofillers to be dispersed uniformly in the matrix, thereby providing a strong interaction between them [8]. A representative SEM image of the P3HT/G nanocomposite is shown in Figure 2. However, certain conditions must be fulfilled, including the use of low-viscosity monomers, a short period of polymerization, and no formation of side products during the process.

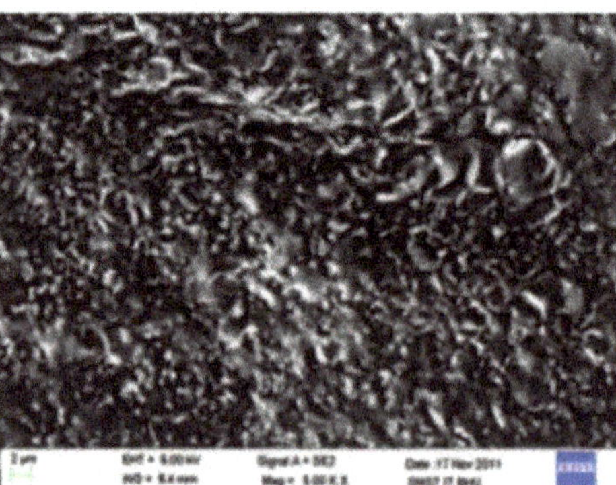

Figure 2. Representative SEM image of P3HT/G nanocomposite (0.1 mg/mL of graphene in P3HT solution). Adapted from ref. [10] with permission from the American Scientific Publisher.

In this work, P3HT/G nanocomposites with different G loadings have been studied. The aim is to assess how the P3HT principal optical properties change in the presence of different G contents. In this sense, the optical bandgap, the hole collection properties and the carrier mobility are studied. In addition, the refractive indexes (n) and the extinction coefficients (k) of the different P3HT/G composites are shown, and the influence of G content on these parameters is discussed.

2. Results

In this section, the changes in the optical bandgap, the hole collection properties, and the carrier mobility as a function of G content are described. Then, the refractive indexes (n) and extinction coefficients (k) of six nanocomposites with different contents of G are graphed. The P3HT/G nanocomposite was synthesized via the in situ oxidative polymerization of the 3-hexylthiophene monomer (3HT) in the presence of graphene. Most of the graphene flakes had 1–4 layers of the size of a few microns.

2.1. Method

The main optical and electronic properties of P3HT were studied before and after the addition of graphene. To complete this, firstly, an exhaustive bibliographic search was carried out. Special attention was paid to results that related to the application of the studied compounds to organic solar cells.

2.2. Electronical Parameters

According to the observations of Chang et al. [2], conductivity increases with increases in graphene loading. Their measurements of conductivity for different graphene loadings are shown in Table 1.

Table 1. Conductivity measurements reported by Chang et al. [2] for different P3HT/G nanocomposites.

Sample	Conductivity (S/m)
P3HT	$9.273 \cdot 10^{-3}$
P3HT/G 0.2 % wt	$1.467 \cdot 10^{-2}$
P3HT/G 1 % wt	$1.549 \cdot 10^{-2}$
P3HT/G 2 % wt	$1.500 \cdot 10^{-2}$
P3HT/G 10 % wt	$1.878 \cdot 10^{-2}$

2.3. Optical Parameters

2.3.1. Bandgap, Hole Collection Properties and Carrier Mobility

The bandgap of P3HT is around 1.9 eV, limiting absorption to wavelengths under 650 nm. In this spectral range, and under the AM 1.5G spectrum, only 22.4% of photons are found. Consequently, decreasing the bandgap leads to an increase in the total amount of photons that can be harvested. However, narrowing the polymeric bandgap results in a reduction in the power conversion efficiency of the cell due to a decrease in the open circuit voltage (V_{oc}). Therefore, a trade-off should be achieved to obtain the optimum bandgap [11].

According to Bkakri et al. [12] and Chang et al. [2], spectroscopic ellipsometry (SE) analysis proves that the insertion of low graphene content in the P3HT matrix reduces the thickness of the film and the optical bandgap of the P3HT/G nanocomposites. As a result, the optical absorption properties of the solar cell increase in the visible range.

According to Saini et al. [3], nonetheless, there is not a clear trend in the P3HT bandgap variation with graphene content. There seems to be a slight increase in the bandgap for low graphene content. However, for high graphene loading levels, both the absolute value of HOMO level and the absolute value of LUMO level slightly increase so that the bandgap remains the same. It would be convenient to clarify this discrepancy between authors regarding the bandgap.

According to Abdul Almosin et al. [13], the use of P3HT/G improves solar cell efficiency due to the enhanced hole collection in P3HT in the presence of graphene. P3HT/G bulk heterojunction prepared by solution processing possesses the advantages of the high carrier mobility of G and the high visible light absorption of P3HT. Che et al. [14] fabricated a phototransistor consisting of a solution-processed P3HT/G bulk heterojunction channel. The device exhibited a hole mobility as high as 3.8 cm^2/(V·s) due to the enhanced charge transport properties of G.

2.3.2. Refractive Index and Extinction Coefficient

In the following section, the evolution of the complex refractive index of P3HT/G with G content is discussed. The data of Saini et al. [3] for n and k for six nanocomposites with different loads of graphene are registered.

Figure 3a shows the refractive index of these six nanocomposites. It can be seen how the n of P3HT with higher loads of graphene is higher, except for the nanocomposite P3HT/G (0.5 wt%), in which n is lower. Regarding k, as is shown in Figure 3b, the extinction coefficient of G/P3HT increases with the load of graphene, except for the case of P3HT/G (0.5 wt%), in which k is lower.

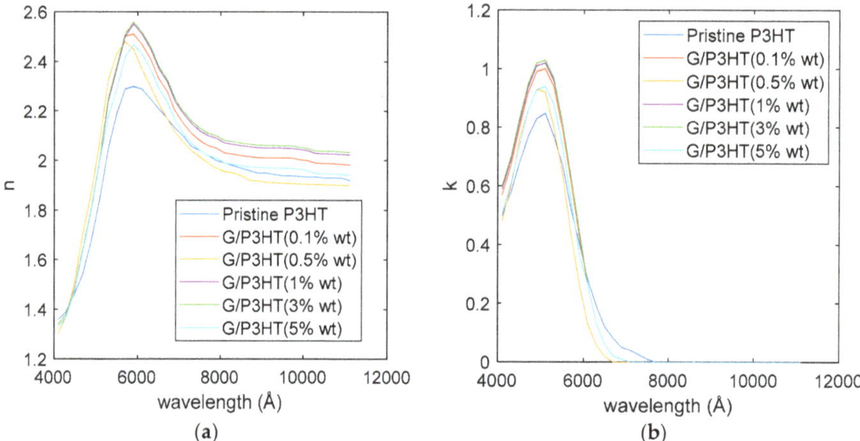

Figure 3. Refractive index (**a**) and extinction coefficient (**b**) obtained from ellipsometer analysis for P3HT/G nanocomposites with different G loads [3].

3. Conclusions

In this work, the influence of the graphene content of G-P3HT on the optical and electrical properties of the compound is assessed.

The addition of different graphene (G) content to P3HT tailors its main optical parameters. In particular, electrical conductivity increases. The hole collection properties and the carrier mobility are enhanced, and the bandgap reduces with increasing graphene content.

It was found that the refractive index of P3HT generally increased with increasing graphene loading; an analogous trend was found for the extinction coefficient of P3HT/G nanocomposites, which increased steadily with increasing graphene loading.

This preliminary study shows the great potential application of these nanocomposites in organic solar cells. Future work will be carried out to further assess their optoelectronic properties in more detail.

Author Contributions: Writing—original draft preparation, L.V.D., R.P.C.; writing—review and editing, R.P.C.; supervision, A.M.D.-P. All authors have read and agreed to the published version of the manuscript.

Funding: Financial support from the Community of Madrid within the framework of the multi-year agreement with the University of Alcalá in the line of action "Stimulus to Excellence for Permanent University Professors", Ref. EPU-INV/2020/012, is gratefully acknowledged.

Institutional Review Board Statement: Not applicable.

Informed Consent Statement: Not applicable.

Data Availability Statement: The data presented in this study are available on request from the corresponding author.

Conflicts of Interest: The authors declare no conflict of interest.

References

1. Díez-Pascual, A.M.; Sánchez, J.A.L.; Capilla, R.P.; Díaz, P.G. Recent Developments in Graphene/Polymer Nanocomposites for Application in Polymer Solar Cells. *Polymers* **2018**, *10*, 217. [CrossRef] [PubMed]
2. Chang, Y.W.; Yu, S.W.; Liu, C.H.; Tsiang, R.C.C. Morphological and Optoelectronic Characteristics of Nanocomposites Comprising Graphene Nanosheets and Poly(3-Hexylthiophene). *J. Nanosci. Nanotechnol.* **2010**, *10*, 6520–6526. [CrossRef] [PubMed]
3. Saini, V.; Abdulrazzaq, O.; Bourdo, S.; Dervishi, E.; Petre, A.; Bairi, V.G.; Mustafa, T.; Schnackenberg, L.; Viswanathan, T.; Biris, A.S. Structural and Optoelectronic Properties of P3HT-Graphene Composites Prepared by in Situ Oxidative Polymerization. *J. Appl. Phys.* **2012**, *112*, 054327. [CrossRef]
4. Ossila. Available online: https://www.ossila.com/products/p3ht (accessed on 7 September 2021).
5. Novoselov, K.S.; Geim, A.K.; Morozov, S.V.; Jiang, D.; Zhang, Y.; Dubonos, S.V.; Grigorieva, I.V.; Firsov, A.A. Electric Field Effect in Atomically Thin Carbon Films. *Science* **2004**, *306*, 666–669. [CrossRef] [PubMed]
6. King, A.; Johnson, G.; Engelberg, D.; Ludwig, W.; Marrow, J. Observations of Intergranular Stress Corrosion Cracking in a Grain-Mapped Polycrystal. *Science* **2008**, *321*, 382–385. [CrossRef] [PubMed]
7. Presto, D.; Song, V.; Boucher, D. P3HT/Graphene Composites Synthesized Using In Situ GRIM Methods. *J. Polym. Sci. Part B Polym. Phys.* **2017**, *55*, 60–76. [CrossRef]
8. Kim, Y.; Kwon, Y.J.; Ryu, S.; Lee, C.J.; Lee, J.U. Preparation of Nanocomposite-Based High Performance Organic Field Effect Transistor via Solution Floating Method and Mechanical Property Evaluation. *Polymers* **2020**, *12*, 1046. [CrossRef] [PubMed]
9. Advani, S.G.; Hsaio, K.-T. *Manufacturing Techniques for Polymer Matrix Composites (PMCs)*; Woodhead Publishing Limited: Cambridge, UK, 2012; ISBN 9780857090676. [CrossRef]
10. Tiwari, S.; Singh, A.K.; Prakash, R. Poly(3-hexylthiophene) (P3HT)/Graphene Nanocomposite Material Based Organic Field Effect Transistor with Enhanced Mobility. *J. Nanosci. Nanotechnol.* **2013**, *13*, 2823–2828. [CrossRef] [PubMed]
11. El-Aasser, M.A. Performance Optimization of Bilayer Organic Photovoltaic Cells. *J. Optoelectron. Adv. Mater.* **2016**, *18*, 775–784.
12. Bkakri, R.; Sayari, A.; Shalaan, E.; Wageh, S.; Al-Ghamdi, A.A.; Bouazizi, A. Effects of the Graphene Doping Level on the Optical and Electrical Properties of ITO/P3HT:Graphene/Au Organic Solar Cells. *Superlattices Microstruct.* **2014**, *76*, 461–471. [CrossRef]
13. Abdulalmohsin, S.; Cui, J.B. Graphene-Enriched P3HT and Porphyrin-Modified ZnO Nanowire Arrays for Hybrid Solar Cell Applications. *J. Phys. Chem. C* **2012**, *116*, 9433–9438. [CrossRef]
14. Che, Y.; Zhang, G.; Zhang, Y.; Cao, X.; Cao, M.; Yu, Y.; Dai, H.; Yao, J. Solution-Processed Graphene Phototransistor Functionalized with P3HT/Graphene Bulk Heterojunction. *Opt. Commun.* **2018**, *425*, 161–165. [CrossRef]

MDPI

St. Alban-Anlage 66

4052 Basel

Switzerland

Tel. +41 61 683 77 34

Fax +41 61 302 89 18

www.mdpi.com

Materials Proceedings Editorial Office

E-mail: materproc@mdpi.com

www.mdpi.com/journal/materproc

9 783036 552873